WAR, CHAOS, AND HISTORY

WAR, CHAOS, AND HISTORY

Roger Beaumont

PRAEGER

Westport, Connecticut
London

Library of Congress Cataloging-in-Publication Data

Beaumont, Roger A.
War, chaos, and history / Roger Beaumont.
p. cm.
Includes bibliographical references and index.
ISBN 0-275-94949-4 (alk. paper)
1. Military history. 2. Chaotic behavior in systems.
3. Complexity (Philosophy). I. Title.
U27.B38 1994
355'.009—dc20 94-19595

British Library Cataloguing in Publication Data is available.

Library of Congress Catalog Card Number: 94-19595
ISBN: 0-275-94949-4

First published in 1994

Praeger Publishers, 88 Post Road West, Westport, CT 06881
An imprint of Greenwood Publishing Group, Inc.

Printed in the United States of America

The paper used in this book complies with the Permanent Paper Standard issued by the National Information Standards Organization (Z39.48-1984).

10 9 8 7 6 5 4 3 2 1

To my mother-in-law
Jean Prentice Naughton
An Elmira alumna, and an amazing lady

and

My much beloved aunt
Angela Dapp Poser
Semper fidelis

Contents

Preface ix

Introduction: Guideposts for Entry into a Strange Land—A Basic Framework and Some Terms xiii

1. Chaos, Complexity, and Defense Analytics 1

2. On the Military Utility of Military History 15

3. The Fragility of Doctrine 34

4. Battle Cruisers, Tank Destroyers, Heavy Fighters: Doctrines Gone Astray 46

5. The Chaotic Sense of War 75

6. The Ordering Impulse 97

7. Matching Frequencies: Complexity and Creativity 113

8. Chaos Upon Chaos: The Environmental Impact of War 135

9. Considerations and Conclusions 145

Select Bibliography 183

Index 191

Preface

At the outset, I must stipulate that contradictions to some things I have set forth in this work may be found in my previous writings, and that I recognize the paradox in relying on history to question the use of it for didactic or prognosticative purposes, at least. Also, in deference to my wife, Penny, who read the manuscript carefully, I concede that I have tended to make heavy use of a fractal mode in relying on strings of nouns and adjectives. As for the conceptual roots of the work, I can only say that many elements fitted into this study came into view long before I knew of the field of chaos-complexity-nonlinearity-ergodics. From the early 1950s to the mid-1960s, my experiences in the army and in business, and the reading of military history raised my awareness of ambiguity and imprecision in historical description and military operations. My service as a military police officer offered special perspectives on the varieties of human behavior, the deficiencies of memory-based testimony, the general tendency in complex organizations to oversimplification, and the prevalence of gaps between rational-linear models and reality.

From 1965 to 1974, further horizons were opened to me at the Center for Advanced Study in Organization, under the mentorship and tutelage of Bernard James, and the many ad hoc faculty in CASOS's various institutes and seminars. There, aside from my administrative duties, under the guidance of Frank Steggert, my task was to develop cases from military history to illustrate principles and effects in phase with the conceptual "three-legged stool" on which CASOS seminars were built: General Systems Theory, organization thought, and creativity. Sailing many figurative seas and entering many strange ports as my career as a historian unfolded from the mid-1960s on, I gathered many beads in the way of concepts, fragments of information, dilemmas, and paradoxes, but they remained unstrung.

I was, for example, often perplexed at the lack of a clear congruence in accounts of the same events among observers and historians alike. Whatever parlor tricks S.L.A. Marshall may have been up to, he did dramatize how great a gap there was between portraying military operations in narrative, maps, and statistics and what actually went on—and between what participants and observers thought they saw and what happened. Another chasm that I long pondered over was that which lay between the selection and preparation of military leaders and what they ultimately did, a subject which, after twenty years at a designated military college and a year at Annapolis, perplexes me still.

A major step toward stringing the conceptual beads came after I presented a paper at the Inter-University Seminar on Armed Forces and Society in 1977 on the dearth of description of command process in military history. After it was rejected by several journals, its publication in the *Naval War College Review* in early 1979 coincided with a major flurry of concern in defense circles over the literal chaos ensuing in a major exercise of command-and-control systems, Operation NIFTY NUGGET, in 1978. The piece drew attention from Admiral Rickover, but more importantly, from Andrew Marshall, Director of Net Assessment in the Pentagon, who invited me to two DoD-sponsored command-and-control workshops in late 1979. At the meeting in Des Plaines, John Sutherland, then a dean at Seton Hall, gave a presentation on the work of Ilya Prigogine, which left me fascinated, uneasy, and baffled.

Although encouraged by Andy and his then-deputy Fred Giessler to move aggressively along the vector of systems, I persisted in sailing a meandering course around the edges of that vast phenomenological continent I had bespied through the fog of my limitations in mathematics. Nevertheless, I found myself increasingly interested in such disparate and seemingly abstruse matters as surprise, unpredictability, the dynamics of wave-front collisions in fluids and gaseous mixtures, and other things that seemed to me more analogous to battle dynamics than the boxes on situation maps. I also had my horizons further widened by close association, from 1979 on, with John Erickson, that most eminent student of Russian *res militariae* at the University of Edinburgh, through the good offices of Andy Marshall, Fred Giessler, and Dick Thomas, director of the Center for Strategic Technology in the Texas Engineering Experiment Station. Especially significant were John's insights on Russian-Soviet control logics and concepts, known as *maskirovka,* and the potential of dynamic modeling with computers, which may very well revolutionize military history over the next generation. His comments on this work were especially useful. Beyond that, Dick's skepticism about the utility of history as a basis for prognostication led me to examine some basic assumptions.

These ideas did not really begin to cohere until the late 1980s, after I had undertaken several projects related to a growing interest in Soviet affairs, which proved highly relevant, given the Russians' excellence in nonlinearity. From 1989 to 1993, four catalysts converged to focus my interest and efforts. First, in 1989–90, I spent a year in the U.S. Naval Academy's History Department as a Secretary of the Navy Fellow, experiencing the most fruitful interaction with colleagues I had experienced since my days at CASOS. There, I discussed systems and chaos matters with Bob Artigiani, who opened my horizons to what was being done in the realm of complexity, especially at the Center for Non-linear Studies. Just as Phyllis Culham apprised me of her seminal work, my son, Eric, sent me a copy of James Gleick's *Chaos,* leading me to pay closer attention to the stream of material in *Scientific American* and other publications.

After I returned to Texas A&M in the fall of 1990, another catalyst came in the form of a comment from Hal Livesay, who observed, after serving on a graduate student's committee in a scientific field, that something was brewing out there in the domain of chaos theory of which historians might well take note. Close upon an encouraging discussion with DuWayne Anderson of the Geology Department, Roy Soncrant, then a graduate student in our history department, learned of my interest in the seminar and provided me with access to his personal collection on complexity, and subsequently his counsel on the manuscript in progress. John Robertson of the Texas A&M Political Science Department also offered his comments. To them, to my daughter Anne for her considerable technical assistance in word processing and enthusiasm for the topic, and to all those whose ideas I have drawn upon, I am grateful—many giants and many shoulders.

Introduction:

Guideposts for Entry into a Strange Land—A Basic Framework and Some Terms

Although most chaos-complexity research has been heavily grounded in mathematics, and researchers have produced a bewildering thicket of equations and named effects, this study is devoid of such apparatus. It is mainly a consideration of the implications of the most essential elements, but does conform to the implicit divergence of nonlinearity—and creativity theory as well—by not following the traditional model of a closed loop of observation-hypothesis-conclusion, but rather by making a series of excursions into figurative halls of mirrors, each unfolding onto galleries of paradoxes and dilemmas that lead, in turn, to further branching. Its three central concepts—war, chaos, and history—can be viewed as "attractors" in complexity theory, pivots that varying forces move around, or as points from which fractal patterns diverge. In either case, the recasting of perspective throws a fresh light on many matters, such as the central paradox that, as adversaries clash in war, each is buffeted, not only by the forces of their foes, but by the colliding of two of the most elemental human urges—one to produce chaos and the other to impose order. That fundamental inconsistency has rarely been in the foreground in the crafting of military history, nor has the chronic confounding of plans and expectations of winners and losers alike in war-fighting, let alone the gap between actuality and historians' attempts to portray it.

Before exploring such paradoxes and considering implications arising from the viewing of war, chaos, and history in light of chaos-complexity theoretics, a setting forth of some basic concepts is in order.[1]

Attractors—points around which "non-linear oscillations," or trajectories, loosely and approximately but nonrepetitively orbit within "stable limits cycles"[2] around ordinary attractors, which are simple and easily comprehended, series of systems, events "settle down," while in the case of strange attractors, ". . . a minute difference in the starting positions of two initially adjacent points leads to totally uncorrelated positions later . . ."[3]

Complexity—a characteristic of systems made up of more than two elements, suggesting intricacy of structure and process, but not randomness, sometimes with a high degree of regularity in their dynamics up to a point of transition; usually implies a reasonable degree of predictability and controllability, which may quickly pass through a state-change into what is or seems to be chaos, such as the effect of a single accident on rush-hour traffic on a freeway, the outbreak of a riot in a crowd or prison, or the "political upheaval in eastern Europe in 1989" flowing out of "long maintained stable states . . . "[4]

Chaos[5]—intricate, turbulent, multicomponent processes, beyond effective monitoring and reasonable approximate depiction or prediction; cannot be precisely duplicated or simulated; has been described as "an order of infinite complexity,"[6] as "a type of randomness that appears in certain physical and biological systems . . . intrinsic to the system rather than caused by outside noise or interference,"[7] and as exceeding "the capacity of a single individual to understand it sufficiently to exercise effective control—regardless of the resources placed at his disposal."[8]

Fractals—sequences of branchings that are increasingly ornate; based on an initial simple format, their replication yields ever-varying but similar configurations, from simple tree-forms to the elaborate Persian rug–like sequences produced by Mandelbrot and Julia sets.

Fuzziness—a state of complexity, not wholly random or chaotic but beyond exact certainty, in which a system's monitor-controller cannot "distinguish between groups of possible outcomes"[9]; used to describe shadings and approximations.

Nonlinearity—a characteristic of complex systems whose dynamics, when expressed as an equation, are "asymmetrically disproportionate,"[10] as opposed to linear systems, in which output is proportional to input.

Sensitivity to Initial Conditions—the propensity of non-linear systems to change properties far out of proportion to the scale of forces changing their original state.

NOTES

1. A very useful chaos-complexity lexicon is A.B. Cambel, *Applied Chaos Theory: A Paradigm for Complexity* (Boston: Academic Press, 1993).

2. See E.C. Zeeman, *Catastrophe Theory: Selected Papers, 1972–1977* (Reading, Mass.: Addison-Wesley, 1977), p. 12.

3. Clifford A. Pickover, *Computers, Patterns, Chaos and Beauty: Graphics from an Unseen World* (New York: St. Martin's, 1990), p. 378.

4. Manfred Schroeder, *Fractals, Chaos, Power Laws* (New York: W.H. Freeman, 1991), p. 34.

5. For a definition, see S.P.R., *Chaos and Creativity* (Joseph Simon, 1984), pp. 2, 8, and 9; for a definition of disaster, see Anthony F.C. Wallace, *Human Behavior in Extreme Situations* [Publication 390, Disaster Study No. 1] (Washington, D.C.: National Research Council—National Academy of Sciences, 1956), p. 13.

6. F. David Peat, *The Philosopher's Stone: Chaos, Synchronicity and the Hidden Order of the World* (New York: Bantam Books, 1991), p. 196.

7. Robert Pool, "Quantum Chaos: Enigma Wrapped in a Mystery," *Science*, 243/4993 (February 17, 1989), p. 893.

8. William L. Livingstone, *The New Plague: Organizations in Complexity* (Bayside, N.Y.: F.E.S. Limited Publishers, 1985), pp. 1–11.

9. Thomas Whalen and Carl Bronn, "Essentials of Decision-making Under Generalized Uncertainty," in Janusz Kacpyrzyk and Mario Federizzi, eds., *Combining Fuzzy Imprecision With Probabilistic Uncertainty in Decision-Making* (Berlin: Springer Verlag, 1988), p. 27.

10. N. Katherine Hayes, *Chaos Bound: Orderly Disorder in Contemporary Literature and Science* (New York: Cornell University Press, 1990), p. 11.

1

Chaos, Complexity, and Defense Analytics

For the last quarter-century, scientists and theoreticians have increasingly focused attention on problems relating to chaos, complexity, randomness, nonlinearity, uncertainty, and turbulence. In those delvings, they have met difficulty in trying to determine how much those qualities were absolutely beyond the limits of exact measurement, rather than being the result of a current inability to perceive and conceptualize certain phenomena that might ultimately prove tractable to such examination. An acceptance of that possibility is seen in the development of concepts that accept such ambiguity, and the use of terms for approximation and intractability like *messy, squishy,* or *sloppy,* and the concept of "fuzzy sets."

A broader sense of chaos is conveyed by the dictionary definitions that associate it with a lack of order and confusion, primordial formlessness, and conditions in which chance dominates. Philosophers and historians long grappled with chaos in trying to make sense of complex events, and in striving to explain and ascribe meaning, while being aware of the uncertainty and unevenness of evidence, and of the imperfections of human perception and memory and the mechanisms used for gathering information. Military historians, like their colleagues in other subdisciplines of history, have the advantages of hindsight, detachment from direct involvement, and the ability to reflect at leisure. On the other hand, they also face unique dilemmas that stem from the nature of war, the most complex, violent, and dramatic of all human collective undertakings, and as a result, one of the most poorly evidenced.

The anxieties and tensions of war, and most especially combat, profoundly affect perception and thought, saturating the senses and blurring perception. Intense and prolonged fear, preoccupation, fatigue, horror, panic, and pain may add to whatever imperfections of human memory normally affect the recasting of events in other categories of history. Beyond that, combatants have little sense of a "big picture," that is, a larger context of their experience, or much inclination or opportunity to record experience. The overall blurring of perception referred to as the "fog of war" includes the direct stresses arising from the ordeal of battle, but also censorship, secrecy, deception, propaganda, camouflage, and rumor. To the extent that perceptions are accepted as real, they may shape the outlook of those relatively few combatants who try to render accounts of their experiences at the time or later.

Beyond all those distorting lenses, the powerful bonding of fighting men may further affect recollection and lead them to align their impressions with the consensus of their comrades and abandon their initial impressions. Military historians' reliance on such distorting lenses and prisms in trying to recapture what has so often been referred to as the chaos of war makes their task of reconstructing warfare extremely difficult, and as some, like de Pawlowski, have argued, "History [military] is only a tissue of fictions and legends . . . merely one form of literary invention and reality counts for very little in the matter."[1] While most veterans and victims of war remain silent or reticent, others have disdained or discounted attempts to reconstruct and analyze combat. Many commanders, including Wellington, Patton, and Eisenhower, have found historical accounts of their campaigns inadequate. An especially strong sense of such limitations has marked the Russian view of military history from Tolstoi through Bekhterev to Soviet "military-historical science." The latter was based on reducing elements of combat process to the gross statistics of the "correlation of forces," while eschewing detailed description and anecdotalism.

Such varying perspectives reflect the frequency with which combatants, witnesses, and students of warfare alike have sensed a great gap between intentions, plans, and hopes on the one hand and on the other hand, what actually happened. The wild disorder of battle has been a main theme in literature touching on war, including Tolstoi's *War and Peace,* Stendhal's *The Charterhouse of Parma,* Crane's *The Red Badge of Courage,* and Mailer's *The Naked and the Dead.* In contrasting the stunning disorder of combat and the tenuous links between its tumult and those trying to control it at a distance with the concise portrayal of it by historians, James Gould Cozzens described battle "[as] a sort of blundering together of forces, most of them in fact out of any rational control—quite different from the after-the-battle maps showing position of troop bodies and their apparently planned movements."[2]

Although many recognize that attempts to accurately depict complex military operations by historians and eyewitnesses alike have been well short of science, military plans and doctrines founded on such accounts have often been been wildly out-of-phase with operational reality. Contrary to Sumner's assertion that "what you prepare for is what you get," those preparing for one thing in war have usually encountered another. While that concept will be examined more closely further along, the basic discrepancy between expectation and realizations touches on the question of how much the chaos of war is chaos per se, or the product of perceptual or mechanistic inadequacies that may be overcome. That in turn comes up against the pertinence of chaos, complexity, and nonlinearity theoretics to the study and waging of war. Some of these theories have already been brought to bear on such specific problems in the military sphere as communications and cryptanalysis, but as yet there is little evidence on whether chaos-linearity might alter military history and analytics, and—collaterally, or beyond those structures—doctrines and methods in the province of warfare itself.

Research on nonlinearity, complexity, and chaos has been analyzed in respect to implications in such diverse fields as fluid mechanics, meteorology, and economics. Those are all areas in which systems' behavior is never exactly duplicated and in which broadly approximate patterns can be seen, but without the precision that would allow for effective prediction of trajectories beyond a broad range. One matter of special interest to researchers in the field is the way in which nonlinear systems, like streams of water, flows of traffic, or rising smoke, "converge to an equilibrium, oscillate stably, or exhibit persistent chaotic behavior within practicable boundaries . . . [in which] very small changes in initial conditions can cause large and unpredictable changes in output."[3] Students of such dynamics have found degrees of order within apparent randomness and complexity that nevertheless follow rules, leading one of them to conclude that "chaotic motion actually consists of an infinite number of unstable periodic motions or orbits" that can be "nudged into regular-periodic motion by continually applying small motions that would push it into one of these orbits. Or conversely, out of order."[4]

Interest in nonlinearity, chaos, and complexity has intensified since the mid-1970s, when some primary nonlinear concepts emerged at the same time that increasing computer capacity allowed theorists to undertake the calculating of nonlinear phenomena that had previously been intractable. A salient construct among those was the result of research meteorologist Edward Lorenz's seeking to understand why major advances in computer modeling of weather systems had not increased the ability to predict climatic conditions, as many had expected. Observing weather systems' "sensitive dependence on initial conditions," he set forth his butterfly metaphor, to dramatize the effect, by suggesting that a weather system's ultimate development might be profoundly influenced in its delicate formative stage by so slight a force as the flutter of a butterfly's wings.[5]

Another basic chaos-nonlinear construct, the fractal, was created in 1975 by Benoit Mandelbrot, whose Mandelbrot Set has been widely used in computers to generate ornate branching patterns to many orders of growth. It has also been employed in analyzing the many divergences seen in nature, such as clouds, snowflakes, the mixing and diffusion of gases and fluids, tree growth, and other noncyclical, nonregular branching phenomena, including organizational structure. As outer boundaries of initially simple fractal geometric figures divide and subdivide repeatedly, each series becomes more complex and seemingly infinite. A clear sense of scale is denied an observer who lacks a reference point. Fractal geometry has been used in computer graphics in representing branching patterns including trees, streams, and biological systems, as well as "iterated maps . . . chaotic flows . . . quantizing chaotic systems . . . disordered materials, and . . . statistical mechanics. . . ."[6]

That spread of patterns provides, analogically, some perspective on the applicability of such concepts to military history and defense analytics, since many military historians have struggled to trace the course of operations through bewildering disorder evident in the documents left in the wake of battle, and, in some cases, the testaments of participants and eyewitnesses. Over the last two centuries, reconstruction and analysis have been compounded as combat has become increasingly dispersed in space, diffuse, intermittent, and irregular, more like the collision of energized particles or interpenetrating wave-fronts, than of lines, blocs, or dense masses. Aside from static warfare and sieges, force arrays became steadily less linear and more unlike their portrayal on military situation maps, or in histories and atlases. Such fluid, asymmetrical, pointillist contact processes and dysphasic movements of opposing forces in battle led S.L.A. Marshall to coin his metaphor of "eddy currents of battle."[7]

The general sense of chaos as total disorder has been altered by the sublevels of order the nonlinear researchers have found lying within the apparent randomness of turbulence, such as the patterns that emerge in what appears to be thorough random mixing. Researchers have been bemused at discovering what has partly been due to overlooking how the symmetrical shape of the container and the mixing device, as well as the relative weight and variance of the substances, present ordering parameters. Consideration of that presents analysts of combat dynamics with the challenge of determining how much chaos in warfare is a fundamental aspect of battle as a process, and truly random, rather than being only apparent—an inchoate disorder many of whose complex patterns are not yet fully perceived or understood, but which might be.

Long before the emergence of nonlinearity, the chaotic nature of war was frequently noted in military history, science,[8] journalism, literature,[9] and even doctrine, as in the 1933 edition of the German army's Field Service Regulations, *Truppenführung*. Four years later, an American naval theorist wrote that we "cannot with certainty predict the complete pattern of the war for which we prepare ourselves."[10] That sense of ambiguity also flavored the view of T.E. Lawrence ("Lawrence of Arabia") on battle dynamics, which he saw as encompassing a spectrum of "fuzzy" elements, including "the possibility of accident . . . [and] some flaw in materials [that] was always in the general's mind," a "felt element in troops, not expressible in figures," all part of "the irrational tenth" of the tactical equation.[11] Those were only points of consternation on a line of analysis that ran back to at least the mid-eighteenth century, along which military theorists and professionals have struggled with problems arising from trying to control ever larger and more dispersed forces moving faster with greater destructive power. Complexity—and chaos—intensified dramatically from the early nineteenth century onward, as a function of the application of steampower, chemistry, electrification, internal combustion and jet engines, and metallurgy.[12] Ironically, those myriad innovations derived from increasingly precise measurements, but in turn eroded the ideals of symmetry and order that had sometimes been closely approached in praxis in the wars of the ancien régime by such maestros as Marshals Vauban and de Saxe.

Plotting those trends toward increased uncertainty and imprecision in battle dynamics along a time line and examining them from the perspective of chaos-complexity theory presents not a pattern of harmonious waves or lines but one of irregular oscillations, reflecting the tension between fractalization and homogeneity in tactical formats. In land warfare initially, from the mid-nineteenth century onward, the increasing of weapons' rates of fire, accuracy, and explosive power, and ever-larger armies induced an abandonment of linearity and coherence as battle formations became dispersed and asymmetrical, diverging into increasingly particularized suborders with the impact of each technical evolution. The polarity of orthodox linearity versus revolutionary diffusion had become widely visible in the late eighteenth century, both in North America and in Europe, on land and at sea.[13] At first, open and diffuse tactics were linked to exotic settings and then to revolutionary thinking and style, whereas mass, precision, and linearity were associated with the forces of reaction. After each of those Classical and Romantic formats led to successes and failures in battle, open tactical formations, light infantry, and column assaults were balanced by the massing of guns in the French revolutionary armies and then by Bonaparte. The two trends merged so that by the end of the Napoleonic Wars, both sides used much the same mix of heavy and light forces. In 1813, the Prussian, British, and Russian armies had all fielded light infantry, artillery, and cavalry, and the balance of elements and tactics in Bonaparte's *Grande Armée* had swung closer to the principle of mass. At sea, after the last great battle at Trafalgar, the pattern of war shifted to more diffuse blockade deployments and convoys, and to a flickering pattern of individual combats in the *guerre de course*.

The Napoleonic Wars were followed by a widespread reversion to geometrical tactical thought and the search for mathematically based methods and principles. Whether a symptom of political reaction, or a tribute to the triumph of Allied linearity over Bonaparte's fluidity at Waterloo, that cycle would recur over the next century. Even though reliance on tactical linearity in land warfare repeatedly yielded defeat and heavy losses, many military professionals clung to symmetry and devised geometrically regular and coherent tactical gestalts. The mowing down by machines guns of the French infantry's wave attacks in 1914 are well known as are those of the British infantry on the Somme. The Germans also suffered heavy losses from close-order tactics in 1914, since some of their military theorists, concerned about a recurrence of the confusion and chaos encountered in the Franco-Prussian War, had proposed returning to close-order tactics to "organize the necessary disorder of the attack."[14]

Although participants and observers alike were shocked again and again by the heavy losses inflicted by such methods, reliance on them persisted through the first half of World War I. The fractalizing of tactical forms in that war yielded mixed results. Open-order tactics reduced infantry losses, but the diffusion of naval power induced by submarine warfare brought Allied merchant ship losses to near catastrophic levels in 1917. Those losses plummeted and German submarine losses mounted when the Allies dispersed their naval power to escort convoys. Aerial warfare followed a reverse curve, as the initial diffuse and sporadic individual combats gave way to more regular clashes of ever-larger masses of fighters and bombers that produced proportionate increases in losses and brought aerial warfare closer to the attrition model on the ground.

Throughout the "Great War" and afterward, many observers and combatants noted a gap in the perception of battle between "the men who really did the job" and the upper command echelons. The essence of that was caught in the report of the British Expeditionary Force's chief-of-staff, Sir Lancelot Kiggell, who burst into tears as he approached the front at Passchendaele for the first time and saw the vast swamps and ponds churned to thick muck by mass barrages that covered the battlefield. When he exclaimed "Good God, did we really send men to fight in that?" his companion, who had been at the front, observed that things were "much worse further up."[15] Yet the military leaders on both sides in World War I were not as heedless to the need for tactical adaptation as has often been suggested since, although they sometimes appeared very slow to learn and react. By early 1918, the front line in France and Flanders had become a web of strongpoints and observation posts, with forward trenches manned in strength only briefly, before assaults or during raids or counterattacks. Other forms of tactical fractalization included the Germans' infiltration and the Allies' tank tactics, the French army's refined infantry tactics and artillery control methods, the creation of a quasi-science of camouflage, and both sides' sophisticated close air-support methods.

Similar patterns of dispersion and fractalization were also evident across the arc of subsidiary "small wars," including the Bolshevik Revolution, the Irish Easter Rebellion and subsequent "Troubles," the Arab Revolt, and von Lettow-Vorbeck's campaigns in East Africa. While there had been many such *kleine Kriege* and insurgencies before, such as the guerrilla warfare in Spain 1807–14, so many microwars erupting or irregulars prevailing over regular forces and linear tactics had not occurred on such a scale all at once. Nor had there been so many major mutinies, or had so many endured their historical aftereffects. That fractalizing and diffusion in both battle and politics set a pattern for other major wars in the twentieth century as boundaries between the military and civilian realms also became less distinct in other ways. Mechanization and electrification continued to extend the range, scale and effect of propaganda, espionage, and international economic systems.

The proliferation of tangential modes of conflict seen throughout the twentieth century, in the World Wars and the many "small wars" of the Cold War/Nuclear Age, conformed to the phenomena of complexity, in the "self-similarity . . . [of] many different processes that have no underlying built-in scales."[16] That was the case in World War II, in which, aside from the Eastern Front, battle was joined less often, and rarely with the grinding intensity of World War I. Over the previous two centuries, from Braddock's defeat onward, the use of compact, symmetrical battle arrays had often yielded heavy losses, but not always. Overwhelming firepower sometimes offset the dangers of massing, as at Omdurman in 1896, in Korea in 1951–53, and in various amphibious assault landings. That persistence of the impulse to employ coherent gestalts sometimes produced grisly absurdities was reflected in the fate of the tandem rows of U.S. battleships at Pearl Harbor, of line-abreast tank charges by British cavalry regiments in Libya in 1941, of the U.S. Eighth Air Force's mass bomber formations over Germany in mid-1943, and of Japanese *banzai* charges.

Many saw such pushing of massed formations into what had become a virtual shredding machine as a product of residual feudalism, or of a fundamental wooden-headedness that denied professional military elites the ability to see the need to disperse forces in irregular formations, and to maintain only tenuous contact with the enemy most of the time, inasmuch as battle was metamorphosizing from a solid to a liquid and, finally, a gaseous state. If the impulse to hold on to linear formats was due to feudal rigidity, there were other vestiges of that in military institutions of European lineage, including styles and attitudes, some of which permeated their successors in former colonies' and client states' armed forces. The tides of modernism penetrated unevenly into the operational and ceremonial levels. Basic organizational structures, rank systems, and forms of socialization, most especially rigid separation of social classes, remained relatively constant, and were often articulately justified.

Although there was some experimenting with alternative models in revolutionary armed forces and in such exotic cases as Sir John Moore's light infantry system, the re-forming of the Prussian army from 1807 to 1813, and Evans Carlson's 2nd Raider Battalion, they did not have a major influence on military organization. Nevertheless, in the mid-nineteenth century, technological evolution seemed to threaten hierarchical authority in the military as it ultimately eroded that of the monarchy and nobility. In stages, however, as in the case of mechanization and electrification, such changes were harnessed to enhance the centralization of command authority. The stratified pyramid remained the basic model of authority in armed forces throughout the world, although by the late 1970s, advances in command-and-control technology led some, both in the Soviet Union and Britain, to consider the dispersion of authority in complex networks. In the United States Army, a prolonged debate raged from Vietnam onward, as junior and middle-grade officers advocated a system of "general mission type orders" (*auftragsbefehl*) in implicit recognition of the chaotic and fragmented nature of conventional warfare.

The issues in that controversy had been identified in Prussia in the late nineteenth century. Applying railways and telegraphy in war had profound effects in a general sense, as they made the raising, deploying, and sustaining of major armies much easier. The rise of the modern general staff was a product of the Prussians' seeking to make that as much of a science as possible and to maintain the myth that the king still commanded the army in the field in war. Although the system had been demonstrably successful in battle, many German officers were concerned that units in combat and high headquarters had not been linked at a reasonable level of effectiveness. Some saw that as good, others did not, and over the next century that gap was steadily narrowed, but not closed; with each narrowing of the gap, results fell short of high expectations. As those contradictory effects became visible in the late nineteenth century, there was no clear consensus among military professionals or analysts on "c-cubed" (command-control-communications) systems' value as "force multipliers" as opposed to creating a false sense of the high echelons' ability to exercise rational, immediate control over a vast range of complex combat dynamics. Toward the turn of the twentieth century, enthusiasm mounted for the Germans' practice of issuing broad orders and letting lower-echelon commanders select their own methods of executing them. Such keenness among the junior officers of several armies, including that of the United States, resembled that of a century later. Yet some German officers were then concerned about the dysphasia that had appeared when microbattles fought by lower level commanders under the principle of *selbstandigkeit* (independence of initiative) warped operations out of alignment with the intent of higher commanders and staffs, especially in placing unanticipated demands on reserves. Some of those Prussian military theorists anticipated the dilemma that unhinged von Schlieffen's grand maneuver scheme in 1914 as they grappled with the tension between "ground truth" and "the big picture." They formulated a "Law of the Situation," but did not resolve the basic quandary, nor could they foresee either the scale or the ramifications of the impending extension of combat in time, space, and velocity on land, at sea, and in the air.

Since new technologies did not bring the processes of battle under precise monitoring and control, some came to view frustration and disorder as inevitable, integral dimensions of warfare. Aside from human emotions, especially fear and anger, the most obvious chaos-inducing forces were increases in range, accuracy, and destructiveness of weapons, faster rates of fire, and growingly sophisticated fire control. Beyond vastly increased damage and disruption of communications, the dispersion and camouflage of forces made reconnaissance and identification of friends and foes alike increasingly difficult. The human costs of advancing along those axes became visible in stages. While the bloody work done by riflemen dug in at New Orleans in 1815 was seen as something of a marginal anomaly, the vast slaughter wrought by refined versions of improved versions of those weapons at Solferino in Italy in 1859 not only made Napoleon III literally ill, but it led to the founding of the Red Cross. Even those battles were not viewed by military professionals as warning posts, nor were the closely spaced bloodbaths of the American Civil War. Many saw the casualties in that struggle as the result of the Union and Confederate forces being masses of overeager and poorly trained volunteers, even though time changed both conditions.

The armies of France and Germany encountered the same dilemmas when waves, lines, and columns of massed infantry collided with concentrated firepower in pitched battles like Gravelotte in 1870–71. Variations on that theme of massed slaughter appeared at Omdurman (1896), at Mukden (1904–05), in literally dozens of the battles of World Wars I and II, in several battles in Korea in 1950–1953, and in Indochina in 1954 and 1968, and most recently, in the 1973 Mideast War, the Iran-Iraq War, and Operation DESERT STORM. Not only did the tendency to concentration persist in ground war, but an adherence to closely spaced bloc and line formations was also evident in some types of land and sea warfare, and a hunger for geometrical coherence continued to shape military theoretics and practice. Time after time, as such patterned arrays of forces and plans crumbled in the swirl of battle, actual control of combat descended to lower echelons.

While devolution and incoherence in control was dealt with unevenly in theory and praxis, major surprise attacks, another variant of chaos-inducing phenomena, increased throughout the twentieth century, including those that were deliberate and those arising from the unanticipated effects of forces on each side energizing in battle as they collided. The dynamic was not the product of physical forces, since surprise derives from experiencing a threat of sudden-change psychic impact—shock, overload, disorientation, or panic—and may therefore not be a direct result of damage, or a product of the designs of a would-be surpriser. Here, consideration of the nonlinearity implicit in Lorenz's "Butterfly Effect" raises another dilemma: the symptoms of strain that lead an apparently stable complex system to suddenly change dramatically as the result of a relatively small stimulus may not be noted or given much weight by commanders and staffs,[17] or by historians afterward. Even if that is true, it is unlikely that taking steps to enhance such sensitivity and the ability to measure systems' states, subtle stimuli, and "pressure points" would be seen as useful by military commanders, policymakers, planners, propagandists and analysts, among others. To make such concessions to chance, fate, and incomprehensibility flies in the face of deeply ingrained predispositions to frame problems and issues in clear and simple terms, and deal with them in a bold and unambiguous manner, or at least to appear to be able to do so.

Even if some aspects of warfare proved tractable to the application of nonlinear concepts in theory or praxis, it would not necessarily reduce chaos in war. As with the "two-edged sword" that Dostoyevsky observed psychology to be, if bringing nonlinear perspectives to bear on diplomacy, war, and politics yielded insights that made waging war more effective, or appear to be, for one or both sides of a quarrel between states,[18] it might yield a new order of complexity. Most importantly, would the possessor of what was seen to be a significant advantage derived from such concepts be more or less likely to resort to war? If it became possible to see chaotic processes, order-states and "double-attractors" previously embedded in the turmoil of combat but undiscernible, such mapping might lead to improved systems of monitoring and control, and help clarify to what extent combat dynamics are chaotic sui generis. If nonlinear analytics helped in discerning the real limits of the ability to bring battle under rational centralized direction, it could set more realistic expectations regarding the influence of command-and-control technology, and help historians gain a clearer view of how confident they should feel in attempting to describe warfare.

To the extent that chaotic phenomena are "a strange type of mathematical order that gives the illusion of being randomness,"[19] if military professionals gained a clearer sense of an order lying beneath apparently random patterns of battle, that would aid the design of deception, or the countering of it, as well as the shaping and arraying of forces, and the timing, phasing, and tempo of maneuvers. The idea that small stimuli at key points might unbalance an apparently highly stable system that is actually highly stressed is of obvious interest to tacticians if it could be harnessed to produce better maneuvers, but nonlinear concepts might also help clarify why warfare has so often slipped out of the control of its initiators and diverged so far from their hopes and intentions. Chaos and nonlinearity might be used to increase the understanding of war dynamics at levels hitherto only sensed vaguely, or wholly out of view, and serve as a practical device for ultimately controlling it or preventing it altogether. At a somewhat more theoretical level, nonlinear concepts might offer military historians and analysts fresh perspectives on the chronic divergence between popular journalistic and official images of warfare versus individual human experiences of war found in diaries, letters or clinical accounts.

Whatever the advantages of applying nonlinear perspectives in various domains might seem to be, some caution is warranted. Not all scientists have agreed as to whether chaos-complexity-nonlinearity is only a slight change of analytical perspective, or a fad, as opposed to being what Thomas Kuhn in *The Structure of Scientific Revolution* called a "paradigm shift," that is, a major, fresh conceptual configuration that moves perspective to a whole new order. Such caution is understandable, in the realm of defense analytics as well as science, considering the mixed results of various concepts that have been applied in military and naval (and air) affairs, including geometry, operational research, Monte Carlo processes, and General Systems and catastrophe theories. Applying complexity to defense analytics might not abstract and dehumanize war any more than, say, war-gaming or systems analysis. It would be lamentable if nonlinearity-complexity-chaos research proved to be only one more fount of jargon and simplicities, a conceptual cul de sac whose initial glitter created a false sense of understanding of an ability to influence complex matters.

Even if such concepts proved to be transient faddism, weighing fresh ideas can lead to a reconsideration of basic principles and assumptions, in keeping with the dictum of the nineteenth-century British military historian and theorist G.F.R. Henderson: "[since] War is turmoil, whatever tends to mitigate confusion, and to make things easier for the fighting man, is of such estimable value that no sanely governed State can afford to dispense with it."[20] Analysts of warfare, historians, and social scientists as well may ponder the debate among scientists over the utility of nonlinear phenomenology versus statistical mechanics in analyzing the "very rare events" that make up their conceptual universe.[21] Analysts of complexity hold that the more rational a paradigm appears to be, the further it lies from the chaos of reality, in keeping with J. Bernard Gilmore's axiom:

> unpredictability, if used as a definitional criterion for randomness, is always a statement about the relative degree of the predictor's knowledge and ignorance, and not a statement about the consistent behavior of the events in question.[22]

Such hypotheses raise special quandaries for military professionals, political leaders, and analysts alike. Is warfare essentially a kind of chaos that cannot be rationally controlled or even understood beyond elemental limits, even in hindsight? Or has it been seen so far as being more chaotic than it really is, and may yet be rendered more discernible and understandable by using better instruments and analytical methods? If chaos is seen as random and not directly predictable, and representable only in approximate terms, whereas complexity, although very complicated, is reducible to a high degree of predictability and precise numerical expression, and such varying states were plotted across a differential spectrum, would it yield something like a signal-to-noise ratio, offering a view of underlying patterns within apparent turmoil? Like the sharpening of focus following the repair of the Hubble telescope, detecting such underlying patterns could help recapture the past with a higher degree of fidelity and derive meaning from analyzing it. Such a chaos-complexity ratio or index would also allow a way to appraise military history and determine the degrees of impressionism, of differentiations between the lurid and fanciful, and literary art forms versus reportage. Other aspects of history, including the nature of evidence, assumptions, and ascription of causation, as will be seen, appear in a different perspective when examined in the light of complexity, chaos, and nonlinearity.

NOTES

1. G. de Palowski, *Dans Les Rides du Front* (Paris: Le Renaissance du Livre, 1917), p. 61, quoted in Jean Norton Cru, *War Books: A Study in Historical Criticism*, Stanley J. Pincetl, Jr., ed. & transl. (San Diego: San Diego State University Press, 1988), p. ix.

2. James Gould Cozzens, *A Time of War: Air Force Diaries and Pentagon Memoirs, 1943–1945* (Columbia, S.C.: Bruccoli Clark, 1984), p. 49.

3. Theodore J. Gordon & David Greenspan, "Chaos and Fractals: New Tools for Technological and Social Forecasting," *Technological Forecasting and Social Change*, 34:1 (October 1988), p. 1.

4. Edward Ott, Celso Grebogi & James Yorke, quoted in Elizabeth Corcoran, "Ordering Chaos," *Scientific American*, 265:7 (August 1991), p. 96.

5. James Gleick, *Chaos: Making a New Science* (New York: Penguin Books, 1987), pp. 11–31.

6. Hartmut Jurgens, Heinz-Otto Peitgen & Dietmar Saupe, "The Language of Fractals," *Scientific American*, 263:2 (August 1990), p. 67.

7. A parallelism is the description of a "chaotic Hamiltonian system" producing "an essentially infinite hierarchy of 'eddies' of varying sizes," Hassan Aref & Gretar Tryggsvason, "Vortex Dynamics of Passive and Active Interfaces," *Proceedings of the Conference on Fronts, Interfaces and Patterns, May 2–6, 1983* (Los Alamos: Center for Non-Linear Studies, 1983), p. 18.

8. E.g., James G. Taylor, "Recent Developments in the Lanchester Theory of Combat," in K.B. Haley, ed., *O.R. '78* (Amsterdam: North Holland Publishing Co., 1979), pp. 773–806; and Charles F. Hawkins, "Modelling the Breakpoint Phenomena," *Signal*, 43:11 (July 1989), pp. 37–41).

9. E.g., Tolstoi's second epilogue to *War and Peace*, Stephen Crane's *Red Badge of Courage*, and the dicta of von Moltke, the Elder regarding the fragility of plans and the ad hoc nature of strategy, e.g., see Helmuth Karl Bernhard von Moltke, *Gedanken von Moltke* (Berlin: Atlantik Verlag, 1941), pp. 9–20.

10. J.C. Wylie, "Why a Sailor Thinks Like a Sailor," *U.S. Naval Institute Proceedings*, 83:8 (August 1937), p. 813.

11. T.E. Lawrence, *Seven Pillars of Wisdom: A Triumph* (Garden City: Doubleday Doran, 1937), p. 193.

12. For an overview of that transition, see Azar Gat, *The Origins of Military Thought from the Enlightenment to Clausewitz* (Oxford: Clarendon, 1989).

13. E.g., Braddock's defeat and the Battle of New Orleans.

14. See Andrews Hilliard Atteridge, *The German Army in War* (London: Methuen, 1915), p. 85.

15. Leon Wolff, *In Flanders Fields* (New York: Viking Press, 1958), pp. 253–254.

16. William Newman, "It's Simply a Matter of Scale," *CNLS Newsletter*, 68:7 (July 1991), January 1991, p. 53.

17. For a discussion of this problem, see Per Bak & Kan Chen, "Self-organized Criticality," *Scientific American*, 264:1 (January 1991), p. 53.

18. For various views on that prospect, see Ronnie Mainieri, "Impressions on Cycling," *Center for Non-Linear Studies Newsletter*, 76:3 (March 1992), p. 1; James Marti, "Choas Might Be the New World Order," *Utne Reader*, 7:6 (November–December 1991), pp. 30–32; John Briggs & F. David Peat, *Turbulent Mirror:*

An Illustrated Guide to Chaos Theory and the Science of Wholeness (New York: Harper & Row, 1989), p. 64; and Siegfried Grossman & Gottfried Mayer-Kress, "Chaos in the International Arms Race," *Nature*, 337 (February 23, 1989), pp. 791–794.

19. Robert Pool, "Is Something Strange About the Weather?," *Science*, 243:4896 (March 10, 1989), p. 1292.

20. G.F.R. Henderson, *The Science of War: A Collection of Essays and Lectures, 1891–1903* (London: Longmans Green, 1912), p. 399.

21. E.g., see Robert Batterman, "Randomness and Probability in Dynamic Theories: On the Proposals of the Progogine School," *Philosophy of Science*, 53:2 (June 1991), pp. 241–261.

22. G. Bernard Gilmore, "Randomness and the Search for PSI," *Journal of Parapyschology*, 54:1 (December 1989), p. 339.

2

On the Military Utility of Military History

Although warfare has often been described as chaotic, many have tried to impose clear-cut patterns on its turbulence, both commanders (and their staffs) in preparing plans and issuing orders, and historians in attempting to describe warfare in hindsight. History has also been drawn upon extensively to shape war plans and doctrines, and in spite of a very mixed record, continues to be highly valued as a trove of potential lessons and models.[1] Not all historians and philosophers, or makers, "consumers," and critics of history have agreed on the validity of using historical cases in supporting argument and criticism along the long track that runs from Herodotus, Thucydides, Tacitus, and Suetonius to Spengler, Toynbee, and Foucault. Until the last half-millennium, such polemics were known only to a literate few who had access to the hand-copied manuscripts. History, like the annals of the Venerable Bede and Froissart, were mainly chronicles, while forgery and the altering of official and historical documents were as widely practiced during the Middle Ages as in Soviet Russia and Nazi Germany. After the movable-type printing presses that appeared in Europe in the fifteenth century were used initially to print classical and religious manuscripts, a flood of original works and broadsides followed as new and old religions laid the foundations of modern propaganda in their struggle to define history, and each used their selected visions of the past to support their claims. Since then, historical argument has been used in academic discourse and to sway large audiences, with varying degrees of concern for authenticity on those who wrote it and those who read it. Although the number of historians and the volume of their output has vastly increased since the mid-eighteenth century, their efforts have been a small fraction of the historical images that have been disseminated by the media and the arts overall.

Until the nineteenth century, literacy remained low and the cost of printed books high, but history had been frequently invoked or distorted to shore up claims to power and control, as in Charles I's insistence on his "divine right" to rule. It was based on allusions to vague, mythical precedents, as were various other disputes over right-of-succession, taxation, and territory. Like catechisms of many faiths, and Magna Carta, such milestones in political thought and social history as the Petition of Right, the Declaration of the Rights of Man, the Declaration of Independence, the Federalist Papers, the Communist Manifesto, *The Thoughts of Chairman Mao,* and *Mein Kampf* were based on allusions to history. As the numbers of historians increased, politicians, ideologues, and diplomats leaned upon their work, which was often presented as something closer to science than literature, to support their assertions. Meanwhile the hunger of millions who had learned to read in the mushrooming of popular education created a mass market for periodicals and books, which in turn fed a mounting interest in the past in the form of histories and historical fiction, which became a vehicle for amusement and edification, but also fed the rising fires of factionalism and nationalism.

As widespread distortion of the past, deliberate and inadvertent, fanned such resentments, some tried to strike a balance, and match history more closely with evidence. But the efforts of those like Von Ranke and others who tried to do that lay far from a full awareness of the intricate forces, social and educational, that shaped their work, such as who controlled their access to sources and to publishability, their own background and education, sense of marketability, political views, their selection of what or what not to study, the acculturation of historians, and the law of location. Such attempts at rationalization and objectivity stood as thin reeds amid the tides of distortion, deliberate and inadvertent, raised by rationalism, Romanticism, and historicism, which all used history as a pit-prop in constructing philosophical, political, and economic arguments. Objectivity and certainty proved rarer than hoped, in keeping with Bacon's view of historical truth as the daughter of time.

Argumentation couched in historical terms, long limited to small groups of intelligentsia, expanded dramatically throughout the nineteenth and twentieth centuries. Increases in literacy, population, and especially the urban middle classes in Europe and North America were paced by a fusion of technical changes in typesetting, papermaking, distribution and marketing systems, and by advances in transport and communications, all of which combined to give the mass media far greater power in disseminating what had once been the rarified intellectual fare of the cognoscenti. Thus the widespread diffusion of the concepts of Marx, Darwin, and Freud intertwined with a vast flow of images and ideas, including fervid symbols and slogans of nationalism, imperialism, modernism, futurism, and pseudoscientific racism. Conveyed in newspapers and periodicals, they fed the rise of new ideological and political movements, and religious controversy as well. Most of those ideologies, along with many cults and fads, were based on a view of history as a process or machine moving along a particular course toward an ideal state, in what John Marcus saw as a search for transcendence, immortality, and redemption in historicity.[2] Struggles for the throttle and steering control of the figurative steamroller of historical evolution led to fierce political battling. Most wars and revolutions of the twentieth century were like the sixteenth and seventeenth centuries' religious struggles, brutal contests of grimly determined elites pursuing visions, based on claims of perceiving purposive patterns amid history's tangled, murky tides, and striving to ride them to destiny or control their flow.

As those figurative maelstroms swirled, their bewildering complexity led some observers to cynicism and others to the extremes of nihilism, or its more subtle variant, existentialism, all tiny rocks as the fast-flowing streams of history were described by others as flowing to one particular grand goal or purpose or another. The majestic, glittering visions of imperial processionals and pageants stretching off over the centuries, if not millennia, were still widely shared in the late nineteenth century, but proved fragile in the twentieth, as cruder conceptual species evolved that enfolded major portions of mankind. Americans' faith in an inexorable march toward "progress" was enhanced by shifts in the international order brought about by technical evolution and the decline of European imperialism that brought the United States to "superpower" status in the wake of World War II. More briefly but intensely, the Nazis blended Germanic militarism and the impulse to conquest with concepts of genetic pruning to produce an ox-breeders' political logic aimed at producing a super-race. With clenched-teeth determination, Marxists, Soviet and otherwise, waded on through sloughs of misery in the faith that the sufferings they endured and inflicted would ultimately yield a pastoral idyll, free of selfishness. As nationalism and ideology overshadowed or assailed established religions, even the most sordid and sleazy caudillos and thuggish dictators wrapped themselves in optimistic slogans and images and proclaimed grandiose visions of history. Although seemingly atavistic, the latter forms endured through the Cold War and, at the end of the twentieth century, showed signs of major resurgence as Marxism faltered and sagged, and capitalism wobbled, but long-dormant coals of the funeral pyre of fascism glowed suddenly brighter in the shifting winds of uncertainty.

Beyond their assumptions of clear trends in history, believers in such systems also often saw those shaping flows working in their favor, as historical tides whose power smoothed and trimmed irregularities and divergences in running toward a better state as a function of what Jacques Maritain called "vectorial laws."[3] That led many, including some historians, to look for patterns, curves, and vectors conforming to the logic of progression established by a particular faith, secular or religious. Such shaping forces, known as "covering laws,"[4] led to a widespread sense, as Isaiah Berlin observed in pointing out how they negated free will and responsibility, "that the world has a direction and is governed by laws," the direction of which "can in some degree be discovered by employing the proper techniques of investigation."[5]

Polarities in the debate had been laid down much earlier. Thomas Carlyle, for example, asserted that history was the aggregrate of biographies of great men, in keeping with Jacob Burckhardt's concept of "irreplacability," while Edward Cheney saw acts of will weighing for nothing against "great cyclic forces."[6] The convoluted logics and ornate intellectual weapons used in such intellectual skirmishes[7] rendered them opaque or made them seem irrelevant to the relatively few nonacademic observers who brought them into view. Although Marxists virtually usurped the field of the philosophy of history by default, it did not diminish the searching for meaning if not wisdom in the past. Propagandists and publicists who drew upon history to shore up their arguments, from Josef Goebbels to Henry Luce, did not flinch from generalizing or abandoning the cautionary guideposts and qualifiers that historians and philosophers used to set boundaries between truth and speculation.[8]

The emotional intensity of such struggles was masked in their academic form, but was clearly visible released in the shrill rhetoric, war, and brutality, evidence that complex forces were at work, far from the level of cognition, or from the intricate and often bizarre intellectualization that suffused various faiths and ideologies, like Friedrich Engels's elegant essays or the murky ramblings of Alfred Rosenberg. While these lay faiths laid claim to the future, or more profoundly, to destiny, many historians strove to stand back from such skirmishes and battles and maintain clinical perspective, focusing on the explanation of specific events or patterns of them. They rejected the utility of history as an aid to prediction,[9] even though the latter lay at the heart of the logic of "covering laws" and such political credos as Marxism, Liberalism, Fascism, and reactionism. However clinical their approach, the historians' assumption of the role of authoritative explainer is something of an effrontery, not only because the tone of certainty was often found to be misaligned with the evidence, but because there is virtually no way to be absolutely sure of that. A nagging sense of such deficiencies has not produced widespread anxiety about the uselessness of analytical history, or the implications of its function as a guide to the future. Rather than reverting to chronicling, most historians have struggled to meld their impressions of complex events into accounts that appear reasonably factual, ascribing meaning by deriving patterns from examining a particular set of, in the parlance of nonlinearity, "boundary conditions." The fact that such contextual matrices are never identical from moment to moment stands in the way of relating a specific case to such configurations, as does their inability to gain full knowledge of what is going on at any one of those moments.[10]

Such quandaries and discontinuities lie far from the view of most "consumers" of history, from those who find it "fascinating" to those who search it for wisdom, useful practical information, or patterns. Structured and deliberate exposure to history, events, and interpretations has shaped the "mind-set" and the behavior of many, from Fenians to Daughters of the American Revolution. In some cases, the influence of historians has been direct and deliberate, when they have written to affect events or policies, or have served as consultants or policymakers. That is potentially paradoxical, since history cannot fully represent actuality. As a result, the use of it—and that has been frequent enough—to bolster authority and arguments, both political and military, may lead to building on a shaky base. The uncertainty arising from that emerges more clearly when military history is observed from a nonlinear perspective, as levels of order hidden within disorder. The immeasurability of complex phenomena and the unique instability of each complex system suggest how military historians in describing the complexities of warfare may depict degrees of order that do not align with chaotic essence, let alone reality.

Reductionism and ordering are as unavoidable in crafting history as in art, and both historians and artists grapple with that fact as they attempt to present their view of a part of the world. The former, however, have not exercised the option of altering their style toward impressionism or surrealism, or, some might argue, super-realism. In any event, attributing causality is an especially thorny problem for military historians as they try to portray the dynamics of warfare, just as it is for peace researchers attempting to fathom the causation, initiation, termination, and prevention of war. However elegant the depiction of battle as a literary exercise, at best only a very broad and rough sketch is possible, even though a confident tone of presentation may suggest a near-exact portrayal of events to some readers, and usually more to those who were involved in the events being described. In spite of that irreducible ambiguity, there is remarkably little of it in the defining of victory and defeat in most battles and wars. Although some struggles ended unambiguously in mass surrender, as at Bataan, Singapore, Tunis, and Stalingrad in World War II, or in collapse or disintegration, like that of Napoleon's forces at Waterloo, or, more rarely, in annihilation, like Custer's demise, or of the Japanese carriers at Midway, the outcome has rarely been so blurred that the labels of victory and defeat did not readily adhere. That chemistry is not fully understood. For example, although Wellington admitted Waterloo was a "damn close run thing," the results were unambiguous. Lundy's Lane in 1814 and Antietam in 1862 were later claimed as victories by both sides, and the winning side at Shiloh and Winchester was afflicted by panic during those battles. Most importantly for historians and those searching history for "lessons," no matter how close the fight, the imaginal power of the immediate perception of victory-and-defeat shapes subsequent analyses. Surviving records and accounts may be found to align with results, or the flow of events weighted—or embellished like Victor Hugo's "sunken road" at Waterloo. Beyond filtering and warping effects of the preferences and style of historians, most primary documents reflecting military operations are the survivors of extensive sifting and winnowing. Those not destroyed or discarded in the tumult of battle or immediately afterward may not survive bureaucratic purging, security screening, or deliberate destruction or alteration. Since materials deposited in archives or other collections are further sorted and ordered, the course of research, organization of data, and interpretation may be shaped by logics and formats that lie far from what initially appeared to be the course of events that they seek to describe and explain—or from what happened.

Although documents are the primary form of evidence of military operations that historians and military professionals depend on in trying to evoke the past, archaeology, even that done at recent battle sites, offers some perspective as well. While that evidentiary dependence is understandable and probably unavoidable, it leads, as in other branches of history, to a degree of reliance that approaches fixation. Although historians in general face problems in judging the validity of evidence, the military historians' task is complicated by the fact that the documents that they rely on, ranging from official records to scribbled notes and memoirs, are fragments and threads of a vast and complex process. While the "paper trail" of certain military-naval bureaucratic processes is dense and well structured, in battle, physical destruction and tumult cut across such linearity. Presuming a certain level of motivation, victors' ability to retain or organize evidence may be no better or worse than losers'. Some, like the Bonapartist and Confederate historians, were especially avid in their writing of history.

At this point, two special dilemmas appear. First, in war, deliberate distortion and inadvertent loss of evidence is greater than in other human activities and experiences, aside from some disasters and conspiracies. Determining authenticity and what may be missing are both forms of deduction, and lie much closer to art than science. Those quandaries, considered with nonlinearity and complexity in view, overlap into another dimension of perplexity. To the extent that warfare is chaotic, minor events and subtle influences that generate disproportionate effects may go unnoted or unrecorded. Since the furor of battle allows very little ordered consideration, weighing, or recording, the relationship between the volume of evidence and the significance of the outcome may actually be inverted. Not only may losers' records be captured intact, but they will not have the opportunity to edit them with the outcome in view, or smooth out imperfections, while victors, especially in the age of *Totalenkrieg*, have often had the power to shape and array evidence to conform with their desires. Not only may the weighing of evidence be shaped by subsequent events, but recording events in the heat of battle has often been a marginal concern to commanders and staffs, or a process to be controlled, restrained, distorted, or suppressed. Beyond that, actual images of battle are very scarce. Even in the age of motion pictures and television, battle "footage" has been fleeting and blurred, and often censored.[11] Advances in literacy and video and audio recording have, in increments, helped offset the dearth of combatants' accounts, but these accounts have often fallen well short of being lucid and exact, especially those rendered long after events, dependent on the comparison of memories with others' or upon the skill of interviewers. Although diaries and letters convey a somewhat sharper sense of events, and especially the chaos of battle as seen by "lower participants," most such writing has been done well away from the scene of action. Throughout the twentieth century, keeping diaries in combat was usually prohibited for security reasons, and letters were widely censored. Beyond that, those writing home usually tried to avoid upsetting recipients. Such filtering and the effect of distance in space and time is evident in the contrast between transcripts of radio and operational messages and orders shaped in the heat of action and those crafted farther away from battle zones, or after the event.

With that in view, it is not surprising that those confronting the fragmented and incomplete puzzle of battle have resorted to an impressionist blending of anecdotes, allusions, generalizations, statistics, maps, and interviews. The obvious riposte to that is, what else could be done? However valid or understandable it may be, if the disdain of many combat veterans toward military history is accepted as valid de facto proof that trying to describe combat accurately and comprehensibly is futile, military historians seem to have little alternative to abandoning their craft or falling back on the broadest of generalizations and mechanistic chronicling. Those are not the only choices. The computer revolution and the growth of networks may offer historians a pathway out of the limits imposed by the dominance of the single researcher-author monograph as the nearly exclusive format in the academic discipline of history, and its journalistic variant as well, but whatever methodologies may emerge, they will be unable to alter the fact that much of the evidence of war is very mechanistic in itself. John Keegan and others have noted how official documents and after-action reports are spare in style and content, flat and unidimensional, aside from some aide-mémoire, intelligence debriefings, court-martial transcripts, and medical records. Those are a relatively small portion of the corpus of military history, and often hard to obtain, leaving historians trying to discern the texture of battle more dependent on interviews, memoirs, diaries, and letters. Not only are the latter relatively scarce and hard to find, but weighing their value as evidence must be considered in view of the limited human capacity to recall stressful experiences, both in a proverbial sense and as measured by psychologists. Leo Tolstoi, S.L.A. Marshall, and Elizabeth Loftus all considered, from different perspectives, the forces that warp recollection, including stress, fear, anger, and anxiety.

Beyond the clouding effect of such emotions akin to passion lies the tendency of individuals to reshape their view of the past to conform with consensus perceptions and to align with their own hopes, expectations, and preferences, or with subsequent events, suggestions, and experiences. Such influences can include journalistic accounts, histories, and memoirs that appear between an event and a serious attempt to recollect and reflect on it, whether after surveying a broad array of evidence, or making a simple reminiscence. To compound the effect, in war, perceptions and memories are shaped or blunted, in battle by a turmoil that dazzles, frightens, befuddles, and disorients as it overwhelms the senses, but also in stressful situations short of direct combat, like flying in bad weather, submarine service, convey escort, medical duties, or intelligence work. Fragmentation of perspective owing to location, security compartmentalization, censorship, and ignorance often denies a full sense of what is happening overall to most of those involved in war, and in an absolute ultimate sense, everyone. It is, then, not surprising that those who share such complex and tumultuous experiences have failed to agree as to what happened. That is visible in the half-century-long debate among those airmen who flew the "Yamamoto mission" over who should have received official credit for shooting down the eminent Japanese admiral Isoroku Yamamoto on April 18, 1943.[12] Such imperfections of memory, aggravated by war's special stresses, present military historians with a task analogous to accident investigators seeking causal patterns from fragments and testimony. Even "small wars" can encompass a dense flow of virtual disasters, and some military operations have resembled air crashes or ship sinkings, with few or no survivors, and without flight recorders, at a remote site that inspectors may never be able to visit at all, or only much later when evidence has diminished or vanished.

The raw physical effect of war on evidence has sometimes been reflected in casualty statistics. At the end of World War II, for example, only a literal handful had survived of the hundreds of "the Few"—the Royal Air Force fighter pilots who flew in the Battle of Britain of 1940–41. Many other subgroups of that conflict were decimated or perished, including much of the U.S. Marine Corps' "Old Breed," four-fifths of Nazi U-boat crews, and nearly all the Japanese kamikazes. Concerted efforts to capture combatants' impressions include interviews in the field by U.S. combat historians in World War II, and the widespread debriefing of airmen after operations. Those have been small cofferdams on a vast river of evidence. Beyond the loss of witnesses and evidence and the reticence of many who survive lies the suppressive effect of technology, over the last century, as telephones, then radios, and most recently computers, each acted to reduce the production of documents reflecting communications, even though increasing overall information flow. (How much facsimile transmissions will reverse that remains to be seen.) Those technologies allowed most armed forces to generate vast amounts of information, much of it seen as marginal, producing what the Russians call *potok* (flood).

That glut has long been a major concern to commanders and designers of battle management networks, but also raises the question of how much of it survives as evidence relative to that lost in the dynamics of battle. Military historians and combatants alike have often sensed that key data and communications did not survive the turmoil of battle, and that events, crucial and minor, were lost to history. Some professionals also doubted that records adequately reflected "backchannel" or informal communications. As a result, it is very difficult if not impossible to judge how much or which gaps in the figurative picture-puzzle of history have been filled in by distortion, surmise, deduction, allusion, or invention. A special problem here is that the training of academic historians strongly discourages them from exploring ranges or alternative hypotheses or "what-iffing" while they search for patterns and for meaning. The human impulse to impose order on randomness, also measured experimentally, can lead to blending, smoothing, and rationalizing which, intentionally or not, may produce distortion.

History, however systematically approached, being an art, can only approximate reality. Since artful melding is what academic historians are trained to do, uncertainties, ambiguities, anomalies, and randomness may be glossed over, discarded, or ascribed another meaning, while evidence that conforms to what seems reasonable and likely is retained or emphasized. As a result, in military history, crucial details that seem minor may be excluded, as the hunger for ordering works against depicting war or battle is bewilderingly complex or chaotic, either as an absolute state, or as sensed by those engaged. The imposing of logic and categories in bureaucratic ordering and record-keeping may determine what is retained or thrown away. Since imposing rationalist templates on investigation and conclusion may also mask complexity and randomness, categorizing may put key anomalies or divergences out of view or obscure the interaction of complex forces by treating them as separate if parallel categories.

While the preference of both military professionals and analysts for sharp, clear images has crowded nonlinearity off to the side, it has rarely pushed it completely out of the picture. The profound disorder of war has not been a primary theme in military history, but treated more like a frieze, or as background noise or static. Although there is something like a realist school in military history,[13] it has, relatively speaking, been a small fringe group, and many who have purported to be realists have lain closer to abstraction, expressionism, and surrealism, but none have come close, analogically, to the deliberate chaos of Jackson Pollock to the degree that novelists have. Nevertheless, offhand and oblique references to chaos and turbulence reflect how often plans and intentions went awry for winners and losers alike. The idea that much lies in the lap of fortune in the exchange of military force—von Clausewitz saw battle as equivalent to cash transactions in business, some dynamics of which also are shrouded in something like metaphysics in economic theory—stands contrary to the image of doctrine and commanders' visions as the sturdy loom on which the tapestry of operational art is woven. A more apt metaphor is the classical image of fate as a blind weaver, but in war, there are many weavers. As a result, nonlinearity raises the dilemma that using history to shape doctrine may widen the gap between the complexity of reality and doctrine, and create an inappropriate sense of order and generate rational expectations, by reducing perception of the actual angles that lie between contingencies and expectations. That is more likely to be the case when doctrine is based on a single or very few historical cases, and on the perception of patterns and methods.

Randomness and disorder in battle have generally been seen by military professionals and historians as fluctuating norms of little practical concern for leaders and planners. Nevertheless, there are some practical dimensions to the problem. A central concept of complexity is that the ultimate effect of a minor influences may not be proportionately slight, the so-called Lorenz or Butterfly Wing[14] effect, noted earlier, which refers to the susceptibility of complex systems, with little or no advance indication, to undergo a sudden, major state-change because of tiny, seemingly trivial forces. This sensitivity of war dynamics to such destabilizing stimuli, deliberate or incidental, makes military historians' tracing of causation, never a simple task, much more difficult. If small perturbing forces, not directly proportional to ultimate effect, are eclipsed or obscured by larger forces in the tumult of battle, they may be dismissed or overlooked, however causationally significant, in keeping with fictional Guards sergeant Bill Nelson's speculation that "you see an empty brass cartridge case lying on the ground . . . and kick it aside without a word. Yet . . . [it] might be the one thing that won the war."[15]

Both materialism and common sense lead to the judging of the significance of forces and factors in warfare on a basis of size, and assume that those forces and factors are measurable. It is not surprising that broad configurations and patterns have been invented by those trying to explain what happened or predict what might happen. What ultimately happens shapes the description of an event and the explanation of how it came to pass, or how it was supposed to. Ideology and politics can come to bear, as in the Duke of Wellington's justification of the British class system in averring that the Battle of Waterloo was won on the playing fields of Eton. Elements and events conforming to expectations, preparations, and preferences are likely to be seen as causationally important, and may even be invented in the process of historical interpretation. The challenge offered by complexity theory here is that faint and subtle forces, including concepts, ideas, accidents, and "turns of fate," like images drawn from military history or popular culture, or personality clashes and minor incidents that leave little or no trace, may disproportionately shape outcomes.

The idea that warfare is a diffuse, unstable aggregate of phenomena subject to tiny, faint, or unseen influences challenges rationalism and linearity, and most military theory and praxis. Paradoxically, nonlinearity also offers a different perspective on the tendency of the "military mind" to obsess over small matters as a kind of intuitive compensation for the importance of tiny details amid the chaos of war. Fixating on irrelevant minutiae and compulsive ordering in military training, socialization and ceremonies may reflect a strong need to perceive and impose patterns amid the turbulence of combat. Battle history portraying coherent flow, pattern, and causation may appeal to the hunger for order, but it conveys a false sense of the essence of warfare. Such order stands in tension with those many accounts of witnesses, victims of war and warriors alike, that convey the impression, even in very small combats, of too much happening all at once to be fully grasped or orchestrated.

Complexities of cause and effect have long been of concern outside the realms of history and philosophy. Concepts aimed at grappling with problems of uncertainty include Bayesian analysis, game theory, fuzzy sets, and classical statistical mechanics. (Tensions between the latter and nonlinear theory were noted earlier.) In the military realm, such complexities are reflected by the classic problem of assessing the exchange of fire between opposing forces, the best-known solution of which is F.W. Lanchester's "square law."[16] In seeking to explain why losses suffered in such exchanges were not a direct arithmetical product of subtracting the smaller from the larger force, Lanchester posited that, all other things being equal, the outcome would be the remainder of subtracting the square of the smaller force from the larger one. It is, of course, widely recognized that outcomes of battles do not conform to such relationships, since all things are never equal.

That fire-exchange model takes on a different cast when examined from the perspective of nonlinearity and fractalization, that is, the fanning out of different elements of the problem to increasing orders of divergence. Close scrutiny of those divisions brings such small elements into view as the differences in gun designs on each side, tolerances, and the wear of each weapon and cartridge. Lying past tactical formations and the phasing and control of fire are "human factors," including leadership qualities as they affect morale, such as individuals' health, morale, training, experience and personality, and beyond those, variants in wind, terrain, and so on. At the very least, the concept of fractalization increases sensitivity to how elusive actual uniformity and a precise measuring of reality is, and how far it lies from "all other things being equal" or from "a level playing field." The consideration of such complexities stands in tension with the high value placed in defense circles on simplicity as a good thing in itself.

Simplicity is listed as a Principle of War, suggesting that complex tactical schemes can go awry, as in the shambles resulting from Washington's attempt to have five columns converge at Germantown. Yet, from the Golden Horde to D-Day, victory has also frequently gone to those employing subtle plans and stratagems, and complex arrays and maneuvers, including elements of force and influences beyond lists of opposing weapons and uniformed combatants. Both the advantage and difficulty of controlling such complexity in conflict was noted a generation ago by systems theorist Ross Ashby, who drew upon the work of Claude Shannon in defining his "Law of Requisite Variety": "only variety can destroy variety."[17] What Ashby had in view was a large, complex system with a "regulator" (e.g., a headquarters or a decisionmaker) whose capacity to monitor and control fell short of utilizing the system's full capacity. He concluded that trying to reorganize toward that goal without technical enhancement would be futile, and that the overall potential of a system, whatever its components, could not exceed the capacity of the regulator element to monitor and control it. A major implication of the "Law" is that an advantage may be gained from presenting foes with many complex possibilities and actions, thus swamping their coping capacity. While the Law of Requisite Variety dealt with complexity rather than chaos, it touched on the quandaries of finding the boundary between extreme complexity and chaos, and how much the former might appear to be the latter to someone with an inadequate capacity to distinguish complexity (ultimately concrete, finite in format, and measurable in spite of diversity) from chaos (diffuse and describable only in general, approximate, and "fuzzy" terms).

That leads to a series of uncertainties. Might complexity be designed to appear chaotic to an observer unaware of its inner logic? Could a complex system, without the intention or knowledge of its designer, contain subtle flaws or imbalances, seeds of chaos that made it vulnerable to sudden transition to a chaotic state, without the potential or degree of imminent transition being apparent? Does war, as a collision of two or more complex systems, yield unfolding orders of chaos, each of which is, in turn, susceptible to minute stimuli? To what extent, either from a combatant's deliberate designs and actions, or from the essential turbulence of war, might inadequate monitoring or processing capacity cause a commander to mistake complexity for chaos—or vice versa? How effectively can command process or historical analysis discern shadings and differences, other than in broad, impressionistic terms such as "fuzzy sets"?

When seen from the standpoint of tactical or strategic thought, distinguishing complexity from chaos is not an abstruse or peripheral problem, for both in warfare and in international affairs in peacetime, such winnowing of seeds from chaff lies at the heart of the intelligence process, and of targeting, threat perception, and tactics. In any of those situations, sensing when a complex system is vulnerable or about to go out of control is vital. Presuming some potential for chaos exists within any complex system, can the roots of it be traced to quantifiable forces or properties? Could a coefficient of chaos in battle be metered—in an operational setting or in hindsight—to determine something like a signal-to-noise ratio? Could its shape, although probably approximate and fluid, be traced? Historical data can be presented graphically and quantitatively, but historians recognize that absolute objectivity, clinicalism, and precision in mapping causation are unattainable ideals. Since writing history is more art than science, as in other arts, the choosing of a subject or point-of-focus for historical analysis is subjective, influenced by human impulse and preference. As irrational as that may seem in academe or journalism, it is not much more scientific a process in other realms, even though dwelling on a point or points in history by military professionals or political leaders might have far greater consequences. The formers' drawing on history for models and perspectives in shaping doctrine and searching for roles and images of war has tended to intensify during long periods of peace when military and naval professionals had little or no experience of war. A special danger there is that concepts and visions formed on that basis may contribute to a sense of war being more tractable to rational design and control than is possible, and lead to unrealistic expectations.

Not only are complexity and chaos obscured in military history and doctrine, but there is also a masking of the strong emotions, great risks, horror, violence, and turbulence of war in military jargon, slang and informal social practices and attitudes; hence the proverbially unemotional language and spare format of doctrine, including the translation of it into detailed procedures and instructions at the operational level in armies, navies and air forces. Beyond the flat and arid nature of documents noted earlier, field manuals, service school curricula, and SOPs (standing operating procedures) serve as pristine administrative valves and conduits that channel and unleash the flow of death and destruction. Semantic insulators lying along the hierarchical line of concept-doctrine-order-action act to blunt and mask realities of combat that surge up the chain of command, and to prevent the emotional or intellectual base for action from flowing downward. For example, few poilus or *landsers* in 1914 would have seen any connection between their plight and the grandly expressed concepts of Grandmaison or von Schlieffen, nor did U.S. bomber crews in 1943 sense any link between Lord Trenchard's influence on Billy Mitchell and the horrors of "Black Week."

When doctrine is viewed as a broad *Ganzfeld* that extends beyond official statements to encompass ideas, cultural values, attitudes, and reflexes, it throws a different light on history. In every country, historical and mythical images, heroes, and events are touchstones of national identity and patriotism—and shapers of military socialization and "mindset." While selectivity and the sanitizing of martial events, heroes, images, and ceremonies usually act to abstract and obscure the carnage and risks of war, those cannot be wholly masked. Nor can anxieties arising from anticipating and preparing for war be fully numbed. The powerful emotions and fears lying beneath the surface of things in military systems make the reliance on unemotional points of reference and simple rote reflexes and procedures more understandable, and the contrasts between military forms and style and the turbulence of war have been examined from time to time.[18] Procedural rigidity, "tribalism," and acting-out seen in training, hazing, and off-duty social activities reflect the tension between the facade of rational control and conformity and deep emotional sub-currents. While linearity and simplicity may offer comfort to those trying to cope with anxieties and apprehensions by numbing, habituating and fixing attention, they also divert attention from the complex aspects of war. Aside from suppressing exploration for alternatives and dulling sensitivity to subtleties, paradoxes and pitfalls, relying on simplification provides an adversary clearer views of stratagem and style. Beyond any disorientation caused by the confounding of linear and simplistic expectations, that also tends to mute the turbulence of war, and also risks augmenting the sense of shock if such plans and hopes are confounded in war. Such rigidity also strengthens the hostile impulses toward higher authority that exist in most hierarchies, and which are heightened in war by the "them-us" tension arising from the clash between group and self preservation and the following of orders. Although it lies at the heart of the command process, that is indirectly addressed in military training, theory and doctrine. With the disorder of combat in view, the Duke of Wellington said that he would as soon describe a ball as a battle. Although in making that assertion he implied that no one could know enough about either one to do so properly, many military historians have tried, and the appraisal of their attempts has been as subjective as their efforts. At its best, harnessing history to shape doctrine is the layering of one linear template upon another, at the risk of generating chaos. Of course, such striving for patterns and meanings offers a way to cope with anxiety as a tension-reducing ritual, as well as reflecting an impulse to impose order upon disorder, just as many judge military history favorably when it has presented the disorderly dynamic of warfare in an orderly way, and in a form pleasing to seekers after linearity and coherence.

The masking or denial of nonlinearity is further enhanced by the lack of a precise gauge-block or index of effect against which military operations can be measured, beyond the crude "bottom line" of victory-and-defeat, or broadly framed impressions of quality, efficiency, or effectiveness. That lack of benchmarks has spawned such incongruities as victors being criticized as much or more than losers. Military history, like other subcategories of retrospection, has sometimes more resembled a legal brief or political argument than clinical analysis. Losers have often been favorably judged because they were seen as being more competent or bolder than the victors, most notably Napoleon's *Grande Armée*, the Army of Northern Virginia, and the Wehrmacht. Winners have sometimes appeared slovenly and disorganized, for example, descriptions of the British Eighth Army in North Africa and U.S. forces in northwest Europe as sloppy, "unmilitary," and so on. Such judgments steer past the turbulence of battle and minimize the possibility that those less fixated on order per se were more closely aligned with function, and thus better able to adapt to war-as-chaos. That may help to explain why writers of fiction and artists have seemed more effective in capturing the randomness and chaos of war than historians, social scientists, and military memoirists. That is not a trivial matter, given the shaping of the historical view of millions in the twentieth century by fictional images of film and television with what looks like history, but is not. History has also been used to shore up claims of rectitude, superiority, and legitimacy by citing past wrongs, shames, or triumphs, and by fabricating or selecting data. Shaping doctrine in military organizations has also sometimes been a kind of political process and has led to concepts being equated with religious dogma and ideology. The effects of formal discipline aside, the tendency to treat the visions and pronouncements of leaders as revealed truth can stifle the sense of the ambiguities and complexities of war, although not always. In some instances, a sense of uniqueness that arises from cohesion, bonding, and esprit de corps as well as individuals' views of their war experience heightens awareness of such diverse particularism. That awareness is visible in some veterans' disdain for military history and in the belief of some of those who have suffered "post-traumatic stress disorder"[19] that resorting to therapy would be futile, since only they or those who immediately shared their particular ordeal could grasp what they endured.

Although there is enough generality in such trauma to make feelings of uniqueness invalid in respect to clinical tractability, each individual's experience is a kind of snowflake. The variety of battle gestalts has increased in modern warfare to include such diverse forms as airborne, commando, bomber and submarine operations, aerial dogfights, and large-scale ground combat (both episodic and sustained) in mountains, deserts, jungles, temperate climes, and so on. In spite of that, military history, perhaps unavoidably, often conveys a sense of commonality by reducing those figurative snowflakes that constitute the storms of combat to the equivalent of the maps, isobars, and tables that meteorologists use to depict blizzards. At this point, considering these problems in light of nonlinear theory raises the question of how much warfare is chaos, as opposed to being very complex, and where the boundaries really lie between details that seem vital in structuring analysis and those that appear extraneous or superfluous. Considering the "fog of war" and the chronic misperception of combat dynamics by commanders and staffs, if a drawing of lines between chaos and complexity showed the former was greater than general wisdom or best professional judgment has deemed it to be, that would conform with the suspicions of Tolstoi and others about military history. Such skepticism is not quite the same thing as seeing history—as many do—as dull, cold, sterile, or mere story-telling, but all such aversions lie apposite to the drawing upon history by military professionals, political leaders, and policy analysts, which presents a dilemma similar to biologists killing to dissect.

Yet what other options have there been? Shapers of military doctrine have fallen back on inspiration, abstraction, and concoction when searching for baselines and models as they drew on history for "lessons." There has been no uniformity of effect as military theorists, planners, and commanders followed hunches, visions, and inspirations, made blind assumptions based on uncertain data, or relied on simulations, games, or other models. Although those were usually based on some perception of history, however loose and impressionistic their ranging was, such harnessing of the past to practicality lies far from the academic ideal of searching carefully for meaningful patterns in the record of the past, without concern for preconception or utility. Admittedly, that, like most other ideals, has lain some distance from practice, considering how often historians have used their skill and knowledge to support ideological, political, and religious arguments.

Military and political leaders and planners have drawn upon history in many different ways, some, like von Schlieffen, delving deeply and carefully into a specific case, and others, like Churchill, drawing images from a broad base of reading and experience. Some historians drawn into the policy process, like Lord Macaulay, had the opportunity to expatiate and offer some depth of view in their counseling; others were asked only to focus on a closely demarcated zone of concern, so narrowly drawn that they had no full sense of the purpose or implications of their involvement, like the "whiz kids" of the Vietnam era, and "beltway bandits" (defense consultants) of the Cold War.[20] The various cases of policy shaping and crisis management that have come to light have reflected how the pressures of urgency in the bureaucratic environment flattens three dimensions to two and drives out the sensitivity to complexity, randomness, and "fuzziness" that exposure to a broader perspective would present.[21] Since those in authority are usually of a practical bent and driven to action, they are inclined to be impatient with attempts to complicate or introduce uncertainty. Those who raise questions that complicate or serve to brake momentum are not likely to be sought for counsel if they do not sense those realities, and conform to organizational and leaders' values and purposes, or learn to speak or write within certain limits of expectation.

Beyond their examining of historical cases and seeking counsel, military professionals are historically "programmed," both formally, in military and other schools, and informally, through experience, reading, and popular culture. While often dismissed as marginal or trivial, the latter has had substantial influence. Popular culture images and broadly based, stylistically well-presented history that conforms to the anticipations and values of the reader may have greater effect than that more carefully crafted but that runs against the grain.[22] Military history remains an art, not a science, and like a medieval pharmacopoeia it is full of worts and elixirs. Some of it is helpful, some noxious, and some innocuous, but much of it has not been tested or even categorized. A salient implication of that with nonlinearity in view is that seekers after nostrums in history should be cautious in appraising the claims of those mixing concoctions, especially since the practice of military art sometimes seems analogically to be closer to alchemy than to chemistry.

If careful study of warfare ultimately proves it to be more chaotic a phenomenon than previously appreciated, or in different ways, that will help explain why so many attempts to measure military performance, quality, and effectiveness precisely, and to plan and predict, have been frustrated, from the elaborate "angle of attack" schemes of the eighteenth century through the bellometrics of the nineteenth, to all the surprise attacks, "soldiers's battles," and "muck-ups" of the twentieth. Some may view it as crashing through an open door to array further proof that warfare is harder to monitor, predict, and control than most commanders, theorists, political leaders, and historians have assumed. Although the vision of war-as-chaos does not lie very far from von Clausewitz's view of war as the "province of chance," whatever their value may prove to be in other respects, nonlinear theories at least provide a mild antidote to overconfidence and wishful thinking by bringing into sharper relief the complexity of warfare that has lain out of view or off to the side in so many accounts. It remains to be seen what the reasonable and absolute limits are of such analysis and theorizing.

NOTES

1. E.g., ". . . the battlefield narrative, the oldest historical form, has a bright and relevant future," David E. Graves, "Naked Truth for the Asking: Twentieth Century Historians and the Battlefield Narrative," p. 52, and "History can save us from making mistakes . . . provide a sound rendering of the reality of the facts . . . separate myth from reality . . . provide . . . perspective . . . depth of understanding . . . leading a policymaker to a better choice than he or she might otherwise make . . . [and] function as a guide," Anne N. Foreman, Foreword, pp. viii–ix, both in David A. Charters, Marc Milner & Brent Wilson, eds., *Military History and the Military Profession* (Westport: Praeger, 1992).

2. John J. Marcus, *Heaven, Hell and History* (New York: Macmillan, 1967), pp. 3, 264, and 267.

3. Jacques Maritain, in Joseph Evans, ed., *On the Philosophy of History* (New York: Charles Scribner's Sons, 1957), p. 77.

4. See Maurice Mandelbaum, "Historical Explanation: The Problem of Covering Law," *History and Theory*, 1:3 (1960), pp. 229–242.

5. Isaiah Berlin, *Four Essays on Liberty* (London: Oxford University Press, 1954), pp. 24–25; also see Berlin's *Historical Inevitability* (London: Oxford University Press, 1955).

6. Jacob Burckhardt, *Reflections on History* (Indianapolis: Liberty Classic, 1975), p. 272; Cheney quoted in Ernest Nagel, "Determinism in History," *Philosophy and Phenomenological Research* 20:3 (1960), p. 291.

7. E.g., gaining understanding through interpretation ("getting the facts straight") is deemed "hermeneutics" in academic circles, cf. William Todd, *History as Applied Science* (Detroit: Wayne State University, 1972), p. 159.

8. E.g., see R.G. Collingwood, *Essay on Metaphysics* (Chicago: Henry Regnery, 1972), pp. 56–57.

9. See William Dray, *Laws and Explanations in History* (Oxford: Oxford University Press, 1957), p. 4.

10. For perspectives on these dilemmas, see Carey B. Joynt & Nicholas Rescher, "The Problem of Uniqueness in History," *History and Theory*, 1:11 (1961), pp. 150–162.

11. For recent critiques of that problem, see George H. Roeder, Jr., *The Censored War: American Visual Experience During World War Two* (New Haven: Yale University Press, 1993).

12. E.g., see *Second Yamamoto Mission Newsletter*, No. 4 (July 1989).

13. E.g., John Ellis, *The Sharp End: The Fighting Man in World War II* (New York: Charles Scribner's, 1980), and Richard Holmes, *Acts of War: The Behavior of Men in Battle* (New York: Free Press, 1985).

14. Derived from the hypothesis that in the very earliest stages of the formation of air masses that evolve into major weather systems, subsequent major variants are affected by such slight influences as the flap of a butterfly's wing. For a discussion, see James Gleick, *Chaos: Making a New Science* (New York: Penguin Books, 1987), pp. 20–27.

15. Gerald Kersh, *Sergeant Nelson of the Guards* (Philadelphia: John C. Winston, 1945), p. 146.

16. F.W. Lanchester, *Aircraft in Warfare: The Dawn of the Fourth Arm* (New York: Appleton, 1916).

17. W. Ross Ashby, *Design for a Brain: The Origin of Adaptive Behavior* (London: Chapman & Hall, 1960), pp. 229–231.

18. E.g., Norman Dixon's *The Psychology of Military Incompetence*, and Alfred Vagts, *A History of Militarism*; a recent example is Raymond Horricks, *Military Mindlessness: An Informal Compendium* (New Brunswick: Transaction Publishers, 1993).

19. For recent perspective from the perspective of veterans, see Ken Scharnberg, "PTSD: The Hidden Casualties," *American Legion Magazine*, December 1993, pp. 14, 16, 18, and 70. For a view of PTSD from the viewpoint of a military historian, see Roger Spiller, "Shell Shock," *American Heritage*, May–June 1990, pp. 75–87.

20. The most incisive critique of that remains David Halberstam, *The Best and the Brightest* (New York: Ballantine Books, 1992).

21. E.g., see Larry Berman, *Planning a Tragedy: The Americanization of the War in Vietnam* (New York: W.W. Norton, 1982).

22. For one analysis of the question, see Roger Beaumont, "Military Fiction and Role: Some Problems and Perspectives," *Military Affairs* 39:2 (April 1975) pp. 69–71.

3

The Fragility of Doctrine

Chaos in human activities has been the theme of many works of art, literature, drama, film, and music, and has underlain belief systems from Stoicism to existentialism. In late 1992, for example, as the "New World Order" was first being racked by powerful cross-currents, newspaper columnist Linda Ellerbee quoted President-elect Bill Clinton's observation that "there is more chaos in the world than there used to be," and suggested that "the chaos and our involvement in it are, I fear, only beginning."[1] In a similar vein of frustration, Frank H. McKnight, considering the many uncertainties of economic prediction, lamented the fact that[2]

> we do not perceive the present as it is and in its totality, nor do we infer the future from the present with any degree of dependability, nor yet do we accurately know the consequences of our own action . . . [and] we do not execute actions in the precise form in which they are imaged and willed.

Economists and defense analysts both face the quandary of trying to bridge theory and praxis as they strive to trace patterns and control vague and shifting patterns amid a complex flow of events that is only partly comprehensible and tractable to influence. Since the costs of making decisions amid that turbulence in both realms are so high, there has been much studying of the past to aid in framing principles and laws, and in shaping plans, policies, and doctrines.

It should be noted that military doctrine is viewed somewhat differently and also in a broader sense than the term "doctrine" is when used in economics and religion, or even generally, as in referring to someone being "doctrinaire." For example, *Webster's New International*, 2nd edition, deemed doctrine to be "1. Teaching, instruction . . . 2. That which is taught . . . put forth as true, and supported by a teacher, a school or a sect," and encompassing "dogma . . . that which is held as established opinion." Dogma and doctrine are widely seen as synonymous, in keeping with *Webster's* definition of doctrine as "a view or opinion as if it were established fact," citing Cardinal Newman's observation that "many a man will live and die upon a dogma; no man will be a martyr for a conclusion."

While terms like "policy" and "doctrine" are often used interchangeably, the former is usually seen as broad guidelines that define what is to be done or not done in a specific area or sphere of interest, and the latter is visualized as lying between policy and detailed plans and operations. Doctrine has been viewed as a great many things, including protocols, guidelines, standard operating procedures, field manuals,[3] and such terse phrases as "No First Use," "The More You Use, the Fewer You Lose," or "Take the High Ground." To a great extent, doctrine serves as a kind of psychic buffer against unavoidable, unpredictable uncertainties and confusions. Although shaped and set forth in a tone of confidence and certainty, and chronically found wanting in modern warfare, doctrine continues to be held in high regard in most military circles. Aside from any value it may have as a security token in the face of uncertainty, it is usually seen as preferable to a lack of doctrine and form in the face of the irreducible ambiguity and uncertainty that confronts military planners and leaders. Nevertheless, some institutions have operated without a fixed doctrine, muddling through like the British Empire, or improvising or "ad hocing" like the U.S. Navy after Pearl Harbor, while others would argue that doctrine is really the amalgam of many rules, logics, values, and reflexes underlying organizational behavior. Just as a lack of policy is a kind of policy *ipso facto*, ambiguous doctrine is doctrine nevertheless.

Doctrine has often been held on to with ferocious tenacity in the face of contrary indications or failure. That is not surprising, given its role in peacetime armed services as an axis of procurement, training, deployment, and operations, and as the formal justification for receiving resources from the political authority. A deeper and less measurable dimension, but which has sometimes had great influence, is the role of doctrine in shaping and bolstering informal social dynamics, and, as a product of them, in defining and maintaining boundaries and hierarchies, and in justifying service, corps, or factional claims of special allocation of status and resources. Since doctrine may be shaped by forces other than formally stated goals or formal agendas, trying to identify and judge the likelihood of contingencies may lead more toward the preferred than the likely, and toward assumptive if not wishful thinking. At the most elemental level, military doctrine is seen as a totem or icon that casts light amid deepening gloom, and to offer a way of overcoming uncertainty and disorder, as well as providing a special edge in the clash of arms.

The tenacity and fervor with which such virtual dogmas are conceived and adhered to stands in clear tension with the near-impossibility of predicting ultimate reality with any degree of precision. Beyond that lies the paradox that intensity of commitment to a particular schema can add to the shock and confusion that comes with the realization that assumptions are out of phase with reality. The more that doctrine creates clear-cut expectations, whether it is based on linear extrapolation from historical cases, peacetime conditions, or wartime experiences, the greater is its potential, when it misaligns with actuality, for increasing the disorientation of those who depend on it as a guide to specific action and a touchstone of emotional certainty. That might also seem to be true when doctrine is simplistic and linear and, as the product of bureaucratic compromise, overlooks consequences beyond the first order, ignores or rejects subtleties and complexities, or labels deviations from group norms as unorthodox. Striving for such generalizations also ignores the dilemma that chaos-complexity-nonlinearity brings into clearer view—that there have been and always will be exceptions to such apparent truisms.

The wild disorderliness and unpredictability of war has been recognized for a very long time. A leitmotif of combat-as-chaos runs through military history, theory, and literature,[4] and has persisted in the face of technical developments that seemed likely to impose rational patterns on the waging of war. For example, Hugh Cole, a U.S. Army official historian, noted how the uncontrolled turbulence of "soldier's battles" persisted in the twentieth century.[5] Other recent attempts to bring chaos theory to bear on war causation and grand strategy and diplomacy respectively include Sue Mansfield's *The Gestalts of War*[6] and Steven R. Mann's "Chaos Theory and Strategic Thought."[7] In his conclusion to *The Censored War*, George Roeder posited that "an understanding of [the Second World War] grounded in study of the experiences it engendered cannot fail to recognize its complexity."[8] Those all suggest that, from a nonlinear perspective, doctrine shaped in peacetime is like a candle burning in calm air that may be suddenly subjected to the blast of a blowtorch in war. The basic challenge presented to shapers of military doctrine by that metaphor is that a sudden transition in boundary states may leap into the domain of turbulence, in keeping with the warning from some students of nonlinearity that "as time passes, unsolved problems within a given paradigm tend to accumulate and to lead to ever-increasing confusion and conflict."[9]

That understanding has been aggravated by the vague ways in which the term "doctrine" has been used by both military professionals and analysts. Dale O. Smith, for instance, defined doctrine as "the philosophy and principles for waging war as held by the military," noting that "there is seldom complete unanimity regarding doctrine." He included such criteria as official support, inclusion in service school curricula, and high level staff acceptance.[10] A more concise view is that of an Israeli officer: "Combat doctrine is the officially accepted view of the nature of an anticipated war."[11] A sharp contrast is a Soviet view of the mid-1980s, of "the military doctrine of a nation" as "the system of views, adopted during a given period, concerning the essence, purposes and character of a possible future war, preparations for it on the part of the nation and its armed forces, as well as the ways of waging it." Drawing upon Lenin's derivation of von Clausewitz, the Soviet author saw military doctrine as a blend of political, technical, and military elements, with "the former obviously dominant."[12] Not too far from that in essence was air power historian I.B. Holley's definition of doctrine as: "what is being taught, i.e., rules of procedures drawn by competent authority[,] . . . precepts, guides to action and suggested methods for solving problems or attaining desired results."[13] His concern for precision contrasted with the implicit tolerance for ambiguity in another American analyst's suggestion that "effective doctrine should be neither as solid as granite nor as shifting as the sands of the desert."[14]

Such ambiguity pervades military analytics, for example, von Moltke the Elder's dictum, well-known in military circles, that "no general plan survives contact in its entirety with the enemy's main body more than twenty-four hours." Whether that was directly derivative from von Clausewitz or not, it is basically consistent with his views of war as the province of chance, and as an instrument of policy, especially when the latter is construed to mean that policy may be bad or good. Although doctrines are generally seen as the first level in the translating of broad goals and policies into military operations, doctrines have often diverged or been wholly decoupled from their initial orientation. In keeping with the sense of chaos theorist David Ruelle's observation that "turbulence is a graveyard of theory,"[15] warfare has been a graveyard of doctrine.

Just as dogma and doctrine have been seen as blending or blurring, so have policy and doctrine. As such broad statements are "translated" to a more literal, specific form and play through to a practical level, the degree of definition in doctrine is more precise than in policy but not as much as in operational plans and orders. In the United States, the meanings of these terms have been further blurred since the early nineteenth century by the labeling of various American presidents' foreign policies, from Monroe to Reagan, as "Doctrines." Looking past the obvious definitional complexities, in democracies—and many dictatorships as well—policy usually takes the form of broad pronouncements and guidelines flowing down from civilian authority onto the armed forces. While military professionals usually substantially influence the shaping of doctrine, forces, tactics, and methods, those selections have not always been made by those who have fought, planned operations, or sustained the fighters, especially in peacetime or at the outset of major conflicts. Considering the unavoidable uncertainties in defense planning and discontinuities that result from the unpredictability of transition from peace to war and back again, Herman Kahn proposed the metaphor of a department store that is closed for years at a time, that may be forced to open at any time without warning and be suddenly deluged by uncertain numbers of patrons, with the store managers lacking foreknowledge of what goods they will demand.

Even though military history is replete with uncertainties and surprises in the coming of war, and twistings and turnings once it has been under way, military institutions have been stereotyped as holding on to useless forms and practices, even when they have yielded diminishing returns or disaster, both in peace and war. Examples include the mounting of sails on American armored cruisers in the late nineteenth century, the advocacy of paddle-wheels versus screw propellers in the Royal Navy in the 1840s, and Japanese "banzai charge" infantry tactics early in World War II. The gap between functionalism and rituals that led Alfred Vagts to draw his distinction between militarism and military effectiveness has also been reflected in the frequency with which weapons have been diverted to uses other than those for which they were intended. As Admiral Hughes noted, by the end of World War II, "no category of warship except minecraft was employed in the U.S. Navy tactically for the purpose for which it had been built."[16] Other such cases include

- the diversion of submarines to commerce-raiding and of destroyers to convoy escort duty in World War I
- employment of heavy fighters designed to be bomber escorts as night fighters, 1940–45
- the use of German 88-mm guns and various U.S. antiaircraft weapons in ground combat in World War II and Korea
- the P-51 *Mustang*, a "hybrid" that resulted from the combining of an underpowered ground attack fighter airframe with a Rolls-Royce *Merlin* engine
- the employment of B-24 *Liberator* bombers against U-boats in the Battle of the Atlantic
- the Soviets' use of Lend-Lease P-39 *Airacobras* as tank-busters
- the firing of white phosphorous shells, originally designed to make smoke screens, against troops in World War II

- the conversion of chemical heavy mortars to an artillery role in World War II and Korea

Those instances present several paradoxes. While military careerists' foresight has often fallen short of the level implicit in the concept of professionalism, the "military mind"—the popular view of military leaders' reflexive rigidity in coping with change—is not a universal trait among them. And contrary to widespread impression that private industry has sometimes been inefficient or greedy, it has also often proven efficient, creative, and adaptable, and has sometimes shown a greater sense of military sensibility than professionals. There are no iron laws regarding any of those. Many cases can be arrayed to support each of those stereotypes—and other cases to puncture them. Soldiers, sailors, and airmen have sometimes improved on poorly designed and manufactured "low-bid" or obsolete systems by improvising, modifying, or redirecting them; civilian inventors and vendors in other instances have compensated for professional military and bureaucratic inertia or lack of vision. An example of the former was the Finnish pilots' skill in the "Winter War" of 1939–40 which offset the shortcomings of the Brewster *Buffalo,* an aircraft type later deemed a "turkey" by Allied pilots in the Pacific campaign in 1941–42. An example of the latter was the plywood de Havilland *Mosquito* I; although rejected initially by several Royal Air Force commanders, after being adopted by the Path Finder Force, it proved the most effective British bomber in World War II, relative to tons delivered and losses, and types of it served in fighter and reconnaissance roles as well.

Understandably enough, such recurring divergences of occurrence versus expectation led many leaders and planners to try to simplify. The high value placed on simplicity in military professional circles stands counter to the profound effect of minor factors on complex processes observed in chaos-complexity research, and to the lack of a pattern of success or failure in war being clearly related to complexity or simplicity. Although General William DePuy posited that successful major operations were characterized by a clear concept shaped by a single author, and that failures resulted from lack of a sharply defined gestalt, the reverse has also been true.[17] Some clearly shaped doctrines and plans proved self-defeating when they led to holding a course against contrary indicators. That was visible in the Japanese Navy's Midway war games, at Stalingrad, and in Goering's ordering Luftwaffe fighters to fly close escort to the bomber formations in the Battle of Britain in 1940.

Such confounding of doctrines and designs suggests how much closer the formulation of them lies to legal argument than to science, or even art, when data is arrayed to support doctrine, or when they are based on narrow assumptions that are founded in turn on dubious statistical adequacy. Where that process becomes gambling may be easier to see in hindsight, but is not always wholly as apparent as in Premier Tojo's waving aside of various technical problems and contingencies and invoking a medieval legend of a miraculous survival following a jump that seemed likely to be suicidal.[18] The events following that deliberate gamble provided a demonstration of how the coming of war was quickly followed by the abandoning or recasting of closely held doctrine, detailed plans, and formats that comprised organizational "holy writ."

In keeping with Marshal Turenne's observation that victorious generals were those who made fewer mistakes, jettisoning or retooling has not always been the rule when doctrine failed. British official historians saw such inertia at work in the Royal Air Force's strategic bombing offensive in World War II:[19]

> Nor were . . . omissions rapidly repaired once the war had begun. On the contrary, the limitations of the force were only gradually discovered, and for more than two years Bomber Command, in spite of a few remarkable successes, was to a great extent lost in the dark, the haze and the searchlight glare.

James Brian O'Quinn offered a corrective to such overreliance on doctrine in his observation that[20]

> strategy deals with the unknowable, not the uncertain. It involves forces of such great numbers, strength and combinatory powers that one cannot predict events in a probabilistic sense. Hence logic dictates that one proceed flexibly and experimentally from broad concepts to specific commitments, making the latter concrete as late as possible in order to narrow the bands of uncertainty and to benefit from the best available information. This is the process of "logical incrementalism."

Some military planners, recognizing the turbulence and damage that will afflict command-and-control systems in war, have sought ways to maintain operations in "degraded states." Shapers of doctrine face an analogy to that problem if they consider how they might operate if their anticipations are confounded. There are both political and conceptual limits to such deliberating. Leaders or planners may not be willing to allow their subordinates to consider negative alternatives, fearing that such turning away from optimism and enthusiasm might weaken morale, by showing hesitancy to superiors or peers. They may also not have the time or the subtlety and insight needed to generate a range of alternatives and analyze them effectively. While most military professionals can cite historical cases in which commanders' obsessive adherence to a doctrine led to defeat, it is not clear how much such dedication added to the degree of damage done by adding to the delay in commanders and staffs reorienting themselves and undercutting the confidence of subordinate forces.

The *Schlieffenplan*'s failure offers some perspective. In that case, especially brilliant planners and much time given to the honing of the concept and method failed to take into account the vast range of probabilities and possibilities. Whether that demonstrates an absolute inability to anticipate all contingencies is a question that the evolution of computers may answer, or it may be found to lead to more rarified and intricate ranges of uncertainty. It any case, it is hard to explain why, in so many cases, both leaders and followers held on to doctrines and plans even after there was good evidence that they were about to or had gone awry. Does that suggest a corollary to the proverb about government, that a doctrine that functions best limits least? Is there a special premium on keeping the proverbial concrete from hardening and setting into firm pathways? Both the shock generated by the failure of doctrine and the strength of the impulse to hold on to it arise from the extremely complex matrix of individual and group behaviors. Some of those that tend to suppress innovation will be noted further on in considering creativity and chaos. Another source of dissonance and potential chaos is the primitive aspect of leadership, one that has had great influence in politics and warfare, that is, the image of leaders as visionaries and prophets. However irrational it may seem—and has been—ascribing to leaders a special sense of the future has been a major dimension of authority, from the "divine right of kings" to the modern concept of the "big picture," and underlies the mechanisms of deference, authority, and privilege of rank.

While some modern wars, especially those of the late seventeenth and early eighteenth centuries, were struggles over boundaries and lines of succession, most major modern conflicts—the Napoleonic Wars, American Civil War, Taiping Rebellion, World Wars I and II, the Chinese Civil War, the Cold War, and Middle East wars—were duels over controlling the shape of the future in a much broader sense. A browse through Cromwell's and Bonaparte's proclamations, *Common Sense, Mein Kampf,* the speeches of Lincoln, Hitler, Churchill, Roosevelt, and Stalin, and *The Thoughts of Chairman Mao* makes that clear enough. Such predictive pronouncements have not been the exclusive product of major leaders—witness the host of religious prophets, seers, pundits, and forecasters—although only some, like Muhammad and Marx, have had an effect proportional to the predictive success-and- failure of political and military leaders.

Obviously, the matrix of power is rendered more intricate by the extent to which political processes, hopes, ideals, and personal visions overlap with doctrine, or when the latter is an expression of the former. But in yet another paradox found in this phenomenological hall of mirrors, the attempt to separate those factors from the shaping of doctrines and strategies has some of the flavor of the biologists' killing to dissect and Hamlet's warnings about the blunting effect of thought on action. Many schemes aimed at gaining advantage in diplomacy, politics, and war have failed, or succeeded only at the first order of effect, but then set up unexpected countercurrents. Trying to sort out the bewildering tapestry of all those elements and the abstracting of it have led to much frustration and consternation. Hitler and Bonaparte, in relying on military tactical doctrine to yield political advantage, both failed see that it was not a pure, separate, and solid ingot, but a fluid alloy that ran together with society, economics, and politics and constantly flowed into new forms. In the Gulf War, many predictions about the likely course of the war by various experts were confounded. That had little consequence, except to the "experts" themselves, but did throw some light on how some had gained substantial reputations by sheer effrontery, and the fact that the foresight of those who actually directed events, however accurate it was, had far more effect in the "real world." It will take some time to gain a fuller perspective on their judgments, but expectations should not be too great, considering how often major leaders' predictive visions have been confounded. It does not seem to matter much. Crediting leaders with special powers to discern the future has led many to attribute great events to conspiracy rather than to ineptitude or simple oversight.

At this point it is useful to render a caveat regarding the role that motive plays in the shaping of history. The purposes of examining the past also range across a spectrum from simple interest and wonder to the making of briefs and the seeking of support for argument and substance for polemic. A substantial number of historians are drawn to the craft so as to gain a platform for expressing strong opinions. While in some cases the consequent immersion in data has led to changes of view and increased clinicalism, that is not universally the case. Since historians are often drawn to a subject by such natural human motives, that adds a substantial degree of complexity to the shaping of history. Those with strong affections and hostilities alike, as they approach judgment of events from a moral perspective, have tended to be selective in presenting the blizzard of information that constantly engulfs those in power outside the focus of their interest. As a result, some may view as significant in hindsight bits of evidence amid the overall data flow that were not salient at the time, as in the historiography of the 1941 Pearl Harbor attack. It is interesting to compare such misappreciations of commanders like Marshal Maurice Gamelin in 1940, and General Walter Short and Admiral Husband Kimmel a year later in

Hawaii, against presentiments expressed earlier, respectively, by Charles de Gaulle and Billy Mitchell.

In spite of the amorphous nature of politics and war, leaders have often tried to appear intuitive, have employed visionary rhetoric, and some, like Girolamo Savonarola, Woodrow Wilson, and Adolf Hitler, felt that they were indeed visionaries. The paradoxes surrounding accuracy in forecasting are recognized at the level of folk culture in the proverb about prophets being without honor in their own land, and in Homer's unhappy Cassandra. Many leaders have survived the failure of their predictions and the devastating consequences that followed; others suffered greatly from minor errors in anticipating events. In 1940, Neville Chamberlain fell instantly into political oblivion when his proclamation in March that "Hitler had missed the bus" was confounded by the massive Nazi attack on May 10. In contrast, Josef Stalin and Franklin Roosevelt both survived politically after each of their armed forces suffered strategic surprise attacks the next year. The contrast between the victorious George Bush and the vanquished Saddam Hussein lies closer at hand, an instance in which the proving out of revised military doctrine did not dovetail with the political dimension. Whatever the specific details of such cases, appraising them lies far from being a science.

What is normative? There is no index or coefficient of prognosticatory effectiveness that can be used to "score" performance. In an off-the-record briefing to Washington-area journalists early in World War II, Admiral Ernest J. King, Chief of Naval Operations and Commander-in-Chief of the U.S. Fleet, made nine substantial misappreciations.[21] However tempting it might be, given King's arrogance, to judge that severely, there is no benchmark to set those calls against, or way to determine where they fell with, the range of predictive ability of U.S. or Allied service chiefs and major commanders at that time. Nor is there any method for determining whether they were more or less competent than they should have been, or than others were who tried to foresee the future at that time. One of King's counterparts, General Henry H. Arnold, the chief of the Army Air Forces, dealt ineffectively with the fighter escort problem, but sensed the potential of the atomic bomb project more clearly than his peers. All the upper-tier American political, military, and diplomatic leaders failed to anticipate events in the Pacific in late 1941, and the armed services failed to modify their doctrine and practice to align with the evolutions of war they observed in Spain, China, and the early stages of World War II.

Pearl Harbor was a surprise in spite of many specific predictions and portents, the immediate evidence buried in a mass of extraneous data. Similar bolts from the blue came in late 1944, as most major U.S. commanders in Europe were stunned by the Nazi stroke through the

Ardennes, and in 1950, in Korea, when General Douglas MacArthur's intelligence chief, General Charles Willoughby, predicted Chinese Communist intervention in the Korean War, then reversed himself just before it happened. There was no clear link between those failures and the fate of the leaders involved. Some of them fell from power, others went on to greatness. Nor is that surprising, since few clear correlations have been drawn between status and competency, however that might be measured in hindsight, for no objective scale is used to measure capacity in military or civilian hierarchies, other than the Darwinian effects of competition, and subjective judgment. A valid system of evaluation would have to encompass anomalies in the turbulence of warfare and politics to a degree well beyond the capacity of current methodologies.

Such tangled ambiguities, along with leaders' attempts at foresight, have shaped or actually taken the form of doctrine, adding layers of complexity-chaos to organizational processes in both war and peace. That leaves historians and military analysts the task of judging whether what seems to be chaos in a specific case is actually an absolute and irreducible property of the phenomenon being observed, a product of the location and impressions of the perceiver, or a result of the methods used to study it. Up to this point, the lack of clear-cut templates and uncertainty about the validity of methodologies have not prevented people from framing doctrines, policies, or plans despite the lack of testable assumptions. Nor does an end to those practices seem close at hand. The persistence of the impulse to fixate on a vision of the future and on certain ways of dealing with it in the face of thin or contrary data, and the strong social forces that reinforce that impulse, make it unlikely that chaos-complexity-nonlinearity research will have much of a braking effect on those tendencies.

NOTES

1. Linda Ellerbee, "And So It Goes: Military Mission," *Bryan Press*, December 17, 1991, p. 3A.

2. Frank H. McKnight, *Risk, Uncertainty and Profit* (Boston: Houghton Mifflin, 1921), p. 202.

3. A recent Russian definition includes "regulations, instructions, recruiting policy and army organization, combat and tactical troops training manuals, morale-building directives, the system of armed forces equipment and the officer training system . . . once officially adopted and announced . . . [it] assumes the form of universally binding regulations . . . [as] the state's fundamental law in the field of defense," cf. Valentin Larionov & Andrei Kokoshin, *Prevention of War: Doctrines, Concepts, Prospects*, trans. Andrei Medved (Moscow: Progress Publishers, 1991), p. 13.

4. E.g., a sense of discordancy and contradiction in warfare runs through Geoffrey Blainey's *The Causes of War* (London: Macmillan, 1973).

5. Hugh Cole, "War's Forgotten Men," *Vignettes*, No. 69, U.S. Army Center for Military History, April 18, 1977.

6. Sue Mansfield, *The Gestalts of War* (New York: Dial Press, 1982), esp. pp. 55–69 and 208–211.

7. Steven R. Mann, "Chaos Theory and Strategic Thought," *Parameters: Journal of the U.S. Army War College*, 22:3 (Autumn 1992), pp. 55–68.

8. George H. Roeder, Jr., *The Censored War: American Visual Experience During World War II* (New Haven: Yale University Press, 1993), p. 155.

9. David Bohm & David F. Peat, *Science, Order and Creativity* (Toronto: Bantam Books, 1987), p. 53.

10. Dale O. Smith, *U.S. Military Doctrine: A Study and Appraisal* (New York: Duell, Sloan & Pearce, 1955), p. 4.

11. Uri Dromi, "The Risks of Doctrinal Stagnation," *IDF Journal*, No. 14 (Spring 1988), p. 24.

12. Genrikh Trofimenko, *The U.S. Military Doctrine* (Moscow: Progress Publishers, 1986), p. 5.

13. I.B. Holley, "Concepts, Doctrines and Principles: Are You Sure You Understand Those Terms?" *Air University Review* 35:4 (July–August 1984), p. 91.

14. Clifford Krieger, "USAF Doctrine: An Enduring Challenge," *Air University Review*, 35:5 (September–October 1985), pp. 16–25.

15. David Ruelle, *Chance and Chaos* (Princeton: Princeton University Press, 1991), p. 52.

16. Wayne P. Hughes, Jr., Fleet Tactics: Theory and Practice (Annapolis: U.S. Naval Institute Press, 1986), pp. 88–89.

17. Examples of victories won in the absence of a gestalt or with a dim conceptual gestalt include the Battle of the Nations, 1814; Waterloo, 1815; the naval battle of Santiago, 1898; and the final battles in France and Flanders, July–December, 1918. Defeats suffered by shapers of clear concepts include Cold Harbor, 1984; the First Marne, 1914; Midway, 1942; Verdun, 1916; Passchendaele, 1917; and Kursk, 1943.

18. In October, 1941, Tojo told Prince Konoye that "it was sometimes necessary for a man to leap from the stage of Kiyomizu," cf. Saburo Hayashi & Alvin Coox, *Kogun: The Japanese Army in the Pacific War* (Quantico: Marine Corps Association, 1959), p. 28.

19. Sir Charles Kingsley Webster & Noble Frankland, *The Strategic Air Offensive Against Germany*, Vol. 1 (London: Her Majesty's Stationery Office, 1961), pp. 297–298.

20. James Brian O'Quinn, "Strategic Change: 'Logical Incrementalism' " *Sloan Management Review*, 20:1 (Spring 1988), p. 24.

21. Glen C.H. Perry, "Admiral King Talks About Allied Strategy, July 1943," ed. Jeffrey G. Barlow, *Pull Together*, 30:2 (Fall–Winter 1991), pp. 1–7.

4

Battle Cruisers, Tank Destroyers, Heavy Fighters:

Doctrines Gone Astray

It may seem ironic that the following historical cases, chosen to illustrate the fragility of doctrine and the chaotic nature of warfare, also point to some pitfalls in using history to formulate doctrine and analyze warfare. That is not the only paradox or irony that will be encountered along the way. One of the best-known major misadventures arising from using history to shape military doctrine is the failure of the major European combatant nations' war plans to gain quick resolution in 1914. The French army's futile and bloody mass assaults along the northeast frontier were a product of the interpretation of the Franco-Prussian War by a "school" of offensive-minded officers. On the other side, the image of an encircling battle of annihilation at the heart of German war plans was based on the rigorous interpretation of a historical event by General Graf von Schlieffen, Chief of the German General Staff in the late nineteenth century, whose strategic plan for fighting France and Russia at the same time was derived from his study of Hannibal's victory over Rome at Cannae. The *Schlieffenplan* had a substantial secondary effect when, in late 1939, it was the basis of the Wehrmacht's initial plan for attacking in the West. Although the concept was abandoned because some key German documents were compromised, when the Nazi assault came in May 1940, Allied deployments and initial maneuvers were shaped in expectation that the Germans would repeat their 1914 game plan. A generation later, during and after the 1962 Cuban Missile Crisis, the *Schlieffenplan* was widely cited as the prime example of a major war plan that got out of control—this as a result of the wide circulation of Barbara Tuchman's *The Guns of August*, a best-selling popular history of the plan.

Those instances suggest how often the past has been searched by those seeking ways to improve military methods, but they do not represent a universal lack of utility. On balance, the results of such attempts to derive and apply lessons and models from history to military practice has been mixed. It has yielded an advantage in some cases, but has led to failure in others. Many cases of it going each way are well known. In the American Civil War, huge casualties on both sides resulted from the deadly combination of rifles and close-order Napoleonic tactical formations. In the Spanish-American War, General William Shafter's perusing of British military history led to his choice of a landing site in Cuba. In the 1920s, tank warfare advocates studied Mongol tactics as they strove to develop command-and-control methods for mobile warfare, and in 1940, the elaborate and costly Maginot Line, built to shield France on the basis of experience in the Battle of Verdun in 1916, failed to deny the Germans a victory.

As with other branches of history, tracking and weighing causal flow in military history is inevitably imprecise, and the attributing of effect is often deductive and subjective, if not impressionistic. Sorting out the influence on the shaping of doctrines of individual impressions and visions versus that of past examples resembles the dilemmas that intellectual and social historians regularly face. Although military doctrines and concepts may be based on historical precedents, in the design of specific methods and technologies or the identification of "lessons" to justify expectations and plans, they are not likely to be the products of broad, objective searching of history for patterns, methods, or general truths. Cases noted previously, like those that follow, reflect an impulse to selective focusing in looking backward on crisis and war. Retrospection ranges from casual browsing, falling back on cases remembered from classroom or professional reading, or searching for solace, a general perspective, or specific nostrums. That reflex was visible in late 1941, when Wehrmacht officers, beset by the most severe winter in Russia in half a century and denied the quick victory many of them had expected, turned to memoirs of Napoleon's 1812 campaign. In the Vietnam era, policymakers defending major U.S. involvement cited as negative precedent the British and French concessions to Hitler in the Munich Crisis of 1938. Later, various critics of U.S. Vietnam policy invoked such past instances as the "Nürnberg Precedent," Allied strategic bombing in Europe in World War II, and the American Revolution.

In considering the following cases, then, the distinction between using history to support a position versus searching the broader record to draw conclusions should be kept in view. Under the influence of strong pressures to focus and simplify that flow through hierarchies, subordinates (including consultants) have often gathered and shaped data to conform to leaders' express wishes, or what are believed to be their preferences. Beyond that, the ongoing, normal balancing of conflicting views and interests and resource allocation in complex organizations—in democracies and dictatorships alike—also affects doctrinal formulation. In the military realm, as in those other provinces, ethics and objectivity have had little effect on the pressures toward norming, focusing, simplification, and conformity to leaders' desires which all shift perspective away from the chaotic essence of battle. Both historical depictions of combat and administrative process may be affected by either deliberate or unconscious attempts to present a degree of order well beyond that which prevailed. Descriptions that convey ambiguity or uncertainty are less likely to be accepted than those presented in a self-confident tone. That is to a great extent due to the fact that leadership, including military command, around the world is based on a general concept of effective central control. Edicts, directives, orders, and plans flow downward from central authority, as leaders strive to impose their vision and will on complex affairs.

Under the assumption of effective central control, tides running to reductionism and convergence in command process, and in shaping military doctrine, have led commanders and staffs to shape formats of force, with very wide, uneven patterns of effectiveness. Some of those schemes failed to take into account operational complexities and turbulence, hence the discontinuity between higher and lower echelons' perceptions found widely in defense analytics, memoirs, and the literature of war. Since such reliance by commanders and staffs on fixed and regular formats and visions stands in tension with the chaotic essence of war, the ability to impose control, that is, to keep events in alignment with a vision amid such uncertainty, is generally viewed as the highest level of practice of the art of war. That high-level commands have rarely been able to impose precise formats and "real-time" control in modern battle is suggested in such metaphors of the "slushy" feeling of control as herding geese, walking in molasses, or steering a supertanker.

It is very difficult to sort out the turbulent nature of war from the degree of competence of commanders and staffs, and the elements of technology, morale, politics, and so on. The myopia and ineptitude of high-command echelons has been assailed by memoirists, journalists, and novelists—and poets and cartoonists as well—and has become stereotyped in the culture at large as the "military mind." That is compounded by military leaders' demeanor of confidence and aggression irrespective of the state of affairs. Although this attitude can have a very positive effect in some settings, it creates a greater angle of discrepancy between such posturing and reality when things go wrong, and there are many indices of that variant. Military professionals' imperfect control over events has been reflected in chronic overestimation of operational effect in modern military operations, usually threefold.[1] The "body count syndrome" in the Vietnam War was only one of many such cases in military history in which data were distorted or censored to various purposes. Not all the dimensions of turbulence in military organizations in war are so easily approximated numerically as casualties suffered and prisoners taken. Perceptions at various levels of organization vary widely, partly due to the inability of armies, navies, and air forces in peacetime to test doctrines and methods under conditions that even remotely approximate the surging emotions, high energy levels, and feverish intensity of combat. How much increasing sophistication in simulation, including virtual reality, can close that gap remains to be seen.

Although those doctrines found wanting in the stunning and disorienting transition from theory to reality in battle have sometimes been adhered to with grim results, they have more often been abandoned or altered to fit unexpected conditions. Here, nonlinearity offers a different view of that disparity in the concept of "attractors"—swirls of trajectories that never replicate themselves, but follow roughly similar but never identical courses around specific points. In essence, that effect can be seen in the debates of politicians, lawyers, theologians, academics, and military theorists, all of whom readily square off over points of difference, form schools of thought, and defend them ferociously, most especially when exact resolution of the question is clearly out of reach. Even when it is obvious that absolute truth does not lie at one point or around one axis, intelligent and otherwise reasonable adversaries fixate on and defend single points. In much the same sense, in military circles, doctrines based on sharp distinctions and fuzzy logic have often attracted fervid support, whereas detached, objective consideration and tolerance for ambiguities is deemed weakness, uncertainty, or "merely academic." In all of those, the aligning of a point of view with reality is subordinated to a perceiver's urge to force events into alignment with visions and purposes, denying apparent reality as a valid constraint. Within that logic, the accurate perception of events is seen as far less significant than the ability to impose one's will upon them.

Although the following cases were chosen to demonstrate the gap between ordered expectations and chaotic realities, they also reflect swirls of doctrinal debate that orbit around the twin attractors of "heavy" versus "light" forces. The cases noted earlier have circuited other polarities, but they too illustrate how history was used to shape or bolster arguments, how narrow focusing obscured contrary evidence, or left patterns out of view that might have been noticed if a broad array had been examined or a particular case had been more thoroughly studied. In crafting military doctrine on the basis of one or a few historical cases, any advantage gained through simplifying also makes the realigning of expectations more difficult for those habituated to those vectors if things go awry.

Since doctrine affects style and action, it allows an adversary a way to identify and exploit the central logic or stratagem pointed to in the defining of doctrines. Beyond Shannon's and Ashby's arguments noted earlier, coping with increasing complexity requires proportionate or greater investment in control systems. Intermeshed with the technical systems involved in such coping are the capacities of individual commanders and staffs, that encompasses a myriad of dimensions and nuances, no less potentially critical for being immeasurable, such as the social limits, formal and informal, on their ability to exercise their capacities, as well as mind-set, culture, and personal attitudes that affect perception, preference, and reflex. Those human dimensions of organizational dynamics generally fall within a fixed range or are treated as a "given" by system designers and analysts, or by historians projecting their own logics and designs. In political history and science, most key actors are dealt with as rational, although there is no sure way to set their specific actions exactly within a range of rationality, let alone to determine individual motives. In war, that assessment is even more difficult, since the unique strains and pressures of war generate special degrees of aberration and irrationality. Nevertheless, behaviors and choices of commanders are usually presented as products of rational choice. How those forces, gross and subtle, affect the shaping of doctrine and war dynamics can be seen in the following cases, each of which demonstrates the wide gaps between expectation in the form of doctrine, the complexities of operational reality, and evasive subcurrents of complexity.

BATTLE CRUISERS

In the late nineteenth and earlier twentieth centuries, naval designers, policymakers, and theorists grappled with a virtual torrent of technical evolutions, including the transition from sail to steam power, the metallurgical revolution that allowed the making of ever bigger guns and thicker armor plate, the change from coal to oil power, the adoption of turbines, and electrification. The arrays of different types of warships that clashed in World War I reflected those transitions, from motor torpedo boats, used with dash and substantial effect by the Italians in the Adriatic, and by the British in the Baltic, to the battleships of almost 30,000 tons that clashed at Jutland. In World War II, a much wider variety of light, fast combat craft was employed. In northwest Europe, the Mediterranean, and the Pacific, British and, later, U.S. light craft engaged German counterparts extensively in the narrow seas, and in the Pacific theater they performed mainly auxiliary shuttle and scouting roles. As the Italian navy resurrected light units that served as the kernel of modern underwater special operations, mass media in World War II, from newsreels to postage stamps, conveyed the mystique of light, fast forces—fighter planes, airborne and commando troops, and frogmen and motor boats—to vast audiences. In spite of dramatic popular images, U.S. Navy PT (patrol torpedo) boats did not fight major warships head-on, as portrayed in the film *They Were Expendable.* Nor did they prove to be a revolution in naval warfare, as envisioned by enthusiasts like General Douglas MacArthur.[2]

A variation on that theme of disparity between image and substance appeared in the form of a line of naval doctrine halfway through the century's first decade—the battle cruiser (BC). The type was conceived by Britain's First Sea Lord Admiral Sir John Fisher ("Jackie") as a light version of the Royal Navy's new *Dreadnought* class, the first "all big gun," heavily armored, fast battleship. The battle cruiser was intended to "outfight what it could not outrun and outrun what it could not outfight." Proposed roles included supporting cruiser sweeps against the German coast, as a blockade or surveillance force, linking cruiser screens to the main battleship line when cruising for a "scrap," serving as a high-speed element of the main battle fleet in a general engagement, shadowing a retiring enemy fleet, and "mopping up" enemy commerce raiders.[3]

The British *Invincible* class battle cruisers were deemed "battleships in disguise," since they gained four extra knots over *Dreadnought*'s twenty-one by carrying eight instead of ten 12-inch guns, and less armor and compartmentation.[4] They were photogenic and dashing, like the PT boats, and like them, their performance fell far short of advocates' claims, since both PTs and BCs were products of fixating on a single element of a variable equation whose full complexity only came into view when war engendered a crossing of the boundary from complexity to chaos. The falsity of the dictum that "speed is armour" was reflected in the loss of thirteen of the thirty battle cruisers built—some 42%, or four times the loss rate of the *Dreadnought* type—bearing out the premonition of a critic who saw the type as a "heavyweight boxer with an eggshell skull."[5]

When the battle cruiser type was pressed upon the reluctant British Admiralty by Fisher, who saw the type as "better than dreadnoughts,"[6] his enthusiasm was shared by the less flamboyant Sir Julian Corbett, a British naval historian and theorist, who thought the BCs' thinner armor could be balanced by increasing watertight compartmentation,[7] a corrective step that the Germans were to take after their BCs took a hammering at Dogger Bank in 1915. Initially skeptical was Winston Churchill, First Lord of the Admiralty, who had tolerated many of Fisher's wild schemes. The speed versus armor problem had also been considered by a U.S. Navy study group in 1904 examining all-big-gun ship feasibility, and they opted for more armor.[8]

The first phase of World War I at sea seemed to validate "Jackie's" claims. British BCs sank two German cruisers in the Heligoland Bight in late August 1914, and three months later, the BCs H.M.S. *Invincible* and *Inflexible* raced to the Falklands to spearhead the destruction of a German cruiser squadron. When, at Dogger Bank in late January 1915, although they missed a chance for a major victory, British BCs sank one of their German counterparts, Fisher's enthusiasm soared, and, on the eve of his retirement, he proclaimed speed to be "everything," urging conversion of two battleships under construction, as well as three new 32-knot super battle cruisers.[9] Those hopes were shattered six months later. At Jutland, May 31–June, 1916, the major sea battle of the First World War, four-fifths of the capital ships lost were battle cruisers—three British and one German. Only 8 crewman survived of the 3,100 aboard the three British BCs sunk.

After the third British battle cruiser blew up, Admiral Sir David Beatty, the Grand Fleet's Battle Cruiser Squadron commander, quipped that "something seems to be wrong with our bloody ships today." Concern about the BCs' flaws endured for a generation. Admiral Sir John Jellicoe, head of the Grand Fleet at Jutland, and Beatty, his successor, blamed their losses at Jutland on thin deck and turret armor. Others attributed them to inadequate damage prevention and limitation, especially the lack of the flash-tight scuttles on magazine doors that the Germans omitted in their ships, and poor damage control organization.[10] Admiral Bacon, for example, argued that "every one of our ships was open to the dread disaster that came to the *Queen Mary, Invincible* and the *Indefatigable.*"[11] The Grand Fleet's heavy losses were more mirrored by the High Seas Fleets' than the comparison of losses in action suggested. *Lutzow,* a total wreck, was later sunk intentionally by German destroyers, while the much-battered *Derfflinger* and *Seydlitz* barely made port.

After half a generation of peace, BC vulnerability came back into view when the sleekest British ship of the type, H.M.S. *Hood,* blew up on May 1, 1941, after the German battleship *Bismarck* quickly found her range. Built before Jutland, *Hood* had not had her deck armor thickened.[12] Many questions remained unanswered. Could any extra increment of speed have really offset the vulnerability presented by the BCs' thin armor? How much was their high loss rate due to their being put into exceptionally risky situations? Whatever the answers, the basic pattern seen in the plight of the BCs also appears in the subsequent cases: they were a linear projection of a complex technology, born of optimistic expectation based on a simplistic vision that was adhered to in the face of skepticism and uncertainty in peacetime, then confounded in war, and finally kept in action far past the point when their utility had became dubious. From the standpoint of complexity theory, while the failure of experts to ascertain the absolute causes of their loss was fostered by the enormity of the explosions, it also reflects again the intractability of such complex structures to analysis which yields causal certainty and the tyranny of small things that are plainly visible in the following cases.

TANK DESTROYERS

In the domain of land warfare, the increasing importance of light infantry and cavalry in tactical theories of the mid-eighteenth century flowed into the wars of the French Revolution and of Bonaparte. That trend toward fractalization in battle methods and forms, as noted earlier, was echoed a century later as various armies created elite shock units late in World War I, such as the French *nettoyers,* German *Sturmtruppen,* and Italian *arditi.* While those specialized forces were disbanded, others were formed between the World Wars, usually to fight guerrillas and insurgents, for example, the Black and Tans, the Spanish Foreign Legion, and the U.S. Army's 29th Infantry Brigade, formed for rapid air-landing operations in Latin America in the early 1930s. Such special light units proliferated during World War II, including the British Commandos, the U.S. Rangers, and the German *Brandenburgers.* Those attempts to pit stealth, speed, surprise, and agility against the growing firepower, mechanization, and armoring of "conventional" ground forces produced mixed results.

As World War II began, a similar trend toward lightness and speed was visible in the domain of mechanized warfare. Like the battle cruiser proponents, armored vehicle designers between the World Wars sought to enhance combat power by increasing speed and mobility rather than mounting heavy armor and guns. That trend, however, was not the product of so clear a vision of gaining advantages by manipulating one or two variables as the battle cruisers had been. Beyond considerations of road and bridge trafficability, logistics, armor and ammunition weight, spare parts, and fuel consumption, tank design was shaped by stringent economies between the World Wars and discussions in the disarmament talks at Geneva of banning tanks. Armored cars and lightly armored tracked vehicles proliferated, including British Bren gun and Carden-Lloyd carriers, and a wide array of French, Japanese, German, and Italian light tanks, which in the U.S. Army were labeled "combat cars." Those types were widely used in the first two years of World War II by the Germans, British, and Italians—and many were destroyed. In 1941, in North Africa, the British army's light combined-arms "Jock columns" did poorly against the German and even the more lightly armed Italian standard divisions, as American light tanks provided to the British under Lend-Lease followed suit. By early 1942, the phasing out of German PzKw Is and IIs and American *Stuart* and *Honey* light tanks reflected a shift toward heavier armor and firepower in all the major combatants' armies, not only on the Russian front but in the Pacific theaters as well. Traces of that early phase remained visible throughout the war, in such vestigial forms as the U.S. M-24 *Chaffee* light tank; German, British, and American armored cars; and the U.S. Army's Tank Destroyer Corps.

The TDC[13] was a reaction to the Nazi panzer victories of 1939–1940. Between the World Wars, U.S. tank development had lagged well behind that of other nations, not only because of minimal funding, but because of poor technical intelligence and a doctrinal precept that "tanks should not be armed to fight other tanks," but should directly support the infantry. In 1940, when American planners matched Allied tank production projections to overestimations of Nazi tank production, it generated a feverish search for an antidote to the panzer hordes that they feared would swamp a landing in Europe. As a result, the Tank Destroyer Corps was formed as a U.S. Army branch in 1941, organized around a doctrine conceived to fill the gap between Allied and Nazi armored strength by forming units of high-velocity, towed guns that would mass quickly against panzer breakthroughs already thinned by screens of entrenched antitank guns farther forward, and on the front line by the infantry and armored divisions' smaller guns.[14]

After the United States entered World War II, the proportion of towed guns in the TDC was reduced, following a revision of doctrine as the result of trials of 37-mm antitank guns placed in jeeps and light trucks during the fall maneuvers of 1941. The success of 75-mm guns in lightly armored half-tracks against Japanese tanks in the Philippines in early 1942 led to a series of tracked TDs that looked like tanks and mounted heavy guns. Their thin armor provided little protection against fire from tank or antitank guns, or from artillery fire. That vulnerability was recognized in the open-top turrets that were fitted on later models that allowed crews a quick exit.

TDC doctrine drifted out of phase with their actual employment in combat. In 1943, in both North Africa and Russia, widespread deployment of antitank guns and mines stifled tank forces' ability to carry out the kind of freewheeling tank maneuvers commonly seen from 1939 to 1941. As a result, all the careful working out of TD doctrine from 1940 to 1944 had little effect on their role in battle. In North Africa, from the Allied landings in November 1942 until February 1943, TD units were assigned such duties as headquarters security and direct support of infantry units, tasks that lay far from their original doctrine. Their sole triumph came at El Guettar, in March, when TDs were massed to counter a major German tank attack. Although they suffered heavy losses, half-track TDs successfully defended U.S. artillery units and destroyed thirty German tanks. In spite of that, subsequent attempts of TDC enthusiasts to convince senior commanders of their potential failed. Lieutenant General George S. Patton, Jr., proclaimed, albeit erroneously, that only a tank could kill a tank, while General Omar Bradley managed to cut TD forces by one-quarter.

Beyond senior commanders' hostility, the TDs encountered new difficulties as the Allied armies crossed from North Africa to Sicily in July 1943 and then invaded Italy in September. Like tanks, their maneuverability was restricted, and their vulnerability to mines and antitank fire was increased by their being channeled in the narrow defiles and raised roads of Italy. That problem was, of course, compounded by their very thin armor. In early 1944, the U.S. Army ended TD doctrinal development, but thirty TD battalions were put ashore in Normandy that summer. During the advance across France toward Germany later in the year, most motorized TDs served as assault guns, while towed types became auxiliary artillery.

Some of the diverging of TD doctrine and practice was the result of the fading of the dire visions that had led to their creation. A basic problem was that the TDC, as a new army branch, lacked the clout of the artillery and armor branches upon whose turf it encroached. Dispersion of TD subunits hampered exchange of information among units, while their TD leaders were junior in rank to commanders of units they supported. As a result, they could not protest when demands were made that took them far from the role envisioned by those who had designed their doctrine. Other factors that eroded their function included the arrival in 1943–44 of bazookas and self-propelled artillery among U.S. forces, and of the former in the hands of German infantry. At the same time, it became apparent to the Allies that Nazi tank production was falling somewhat short of the levels projected early in the war, and the increasing bite of Allied air power, both close air support and strategic bombing, was affecting not only production but the whole flow process, including transportation, supply and repair, and deployment in battle. Beyond that, the bulk of the Third Reich's armor had been deployed on the Eastern front, where much of it had been destroyed, while defense against the Allied bomber offensive had absorbed about a third of Nazi gun and optical production.

By the end of the war, tank destroyers were serving mainly as motorized artillery, or as feeble variants of the assault guns that had become a mainstay of the Wehrmacht and the Red Army. Even though they had been used in tasks far from the visions of their founders, at the end of the war it was estimated that TDs inflicted twice the casualties on German tank forces as U.S. infantry divisions did, at one-fifth the loss rate. At that point, it did not matter. In 1946, a General Board's critical findings led to the disbanding of the Tank Destroyer Corps, and a generation later the TDC was seen as an anomaly. The shortest-lived combat branch in United States Army history, it was seen to have been "half-way between artillery and tanks" and "had unnecessarily complicated" armored vehicle development and obscured "the need for arming tanks with adequate guns."[15] Beyond that, its brief history provided a stark example of the erosion of doctrine by operational necessity.

HEAVY FIGHTERS

In 1959, General Ira Eaker, who had commanded the Eighth Air Force in 1942–43, argued that the U.S. Army Air Force had lacked long-range escort fighters at that crucial point in World War II because such an aircraft type was beyond the state of the art at that time. He asserted that planning a "conflict or strategy with weapons which couldn't be produced for several years . . . would be stupid," on the grounds that "nobody here, or abroad, or even in Germany, which was ahead of us in fighter development, even suggested the possibility of long-range fighters."[16] Eaker's memory served him ill. Several long-range fighter types were developed and deployed by major air forces in the late 1930s, since the need for escort fighters had been foreseen by several air power theorists who envisioned mass bomber attacks colliding with heavy antiaircraft and fighter defenses.[17] Nor had the heavy losses of unescorted bombers on several occasions in Spain and China gone wholly unnoted by U.S. air planners and leaders.[18]

In the late 1930s and early 1940s, the U.S. Army Air Service, the Army Air Corps, and Army Air Forces presented a series of proposals to the U.S. aviation industry for long-range escort fighters, a half-dozen of which reached at least the prototype stage.[19] By early 1944, modification through additional internal tankage and the use of drop-tanks had extended the range of the Lockheed P-38 LIGHTNING, the Republic P-47 THUNDERBOLT, and the P-51 MUSTANG enough to allow them to act as daylight bomber escorts, and later that year, with some effect, against German jet and rocket aircraft. The discomfiture beneath Eaker's defensiveness was born of the losses incurred in mid-1943 in pressing home massed daylight attacks against stout German air defenses before those assets were in hand. The unexpectedly high cost was compounded by the results, far less substantial than envisioned in the doctrine that set forth how they were to be used.

The need for escorts came in and out of the focus of U.S. air leaders from 1933 to 1943, partly because of technical developments, and partly because of turbulence in mobilizing the U.S. economy for war production, followed by a further surge after Pearl Harbor. Bomber and fighter aircraft speeds had come very close together in the late 1920s and early 1930s, but then diverged sharply in favor of fighters in the late 1930s. In 1931, the standard U.S. pursuit (fighter) plane, the P-26, flew about as fast as the new B-10 medium bomber, but fighters' speed had nearly doubled that of bombers by the decade's end. Initially, the increase in velocity yielded little advantage, since the high fuel consumption of all-metal fighters carrying ever-heavier loads of guns and ammunition sharply limited their time aloft. Beyond that, radar was in its infancy, and relatively crude sound and visual detection systems forced dispersion of fighters and antiaircraft guns to cover eventualities.

Discounting the darker implications of projecting the intersection of those trends, many airmen maintained a view that massed strategic bombers would win out handily in a battle with air defenses, in keeping with a scenario devised by the Italian air power theorist General Giulio Douhet in the early 1920s.[20] The actual influence of Douhet's visions on U.S. air doctrine is unclear, but they did not different widely from those of British prime minister Stanley Baldwin, who proclaimed in 1932 that "the bomber will always get through." Not all U.S. Army Air Corps bomber advocates were so wildly optimistic, however. From the mid-1920s to the mid-1930s, U.S. Army Air Tactical School bombardment manuals anticipated heavy losses for unescorted bomber formations,[21] although such concern was less evident at higher command levels. For example, in 1931, a U.S. Army Air Corps maneuver umpire's report concluded that "increased speed and limitless space" would allow bombers to reach targets relatively unscathed, and that it was therefore "inconsistent with the employment of air forces to develop fighters."[22]

As variables in the tactical equation increased, debate on that issue between the fighter and bomber factions in the Air Corps intensified, the latter claiming that the all-metal B-10 bomber and Norden Mark XV bombsight would make fighter defenses ephemeral, much as Douhet predicted in the early 1920s. Citing the short range of fighters, bomber enthusiasts envisioned their massed formations would have only momentary contact with them, and like many airmen in Europe, feared antiaircraft guns more than fighters until well into World War II. In 1933, a sharp clash on these questions erupted following maneuvers overseen by General Oscar Westover. Captain Claire Chennault, later commander of the American Volunteer Group ("Flying Tigers") and of the Fourteenth Air Force in China, wrote an eight-page riposte to an after-action report authored by Lt. Col. Henry H. Arnold, later Chief of the Army Air Corps, 1938–41, and Commanding General of the U.S. Army Air Forces in World War II.[23] Swimming against that tide of mounting enthusiasm for unescorted bomber formations, Chennault authored a series of articles that appeared in the *Coast Artillery Journal,* which although off the mark in some respects, called for escort fighters to protect bombers.[24]

The fighter-bomber debate was also reflected in hearings before the Federal Aviation (Howell) Committee in 1934. Of six Air Corps officers testifying, five were bomber advocates, and one was a War Plans Division representative from the Army General Staff, whom Chennault baited. After being dropped from the Command and General Staff College list, Chennault resigned on medical grounds.[25] At the same time, the Air Corps Tactical School's bombardment text reflected a mounting enthusiasm in anticipation of the delivery of the first U.S. mass-produced, four-engined, all-metal bomber, the Boeing B-17 FLYING FORTRESS. The B-17s' speed, range, and array of machine guns generated widespread confidence that they would brush past interceptors and be able to bomb with great accuracy out of the range of antiaircraft fire. Yet a note of caution persisted, in the form of a recognition that experience in the air wars of the 1930s might dictate a need for escort fighters.[26] Strategic bombing enthusiasts were also constrained by the fact that building long-range bombers that could reach countries outside the Americas was politically unpopular in the pacifist climate of the 1920s and '30s. As a result, the B-17 was designed to meet Air Corps' specifications as a "coast defense" aircraft, and dovetailed with the defensive "hemispheric defense" strategy.

Since a detailed plan for the strategic bombing of Germany was not drawn up until the late summer of 1941, Arnold's claim in his memoirs after the war that he had foreseen that the lack of escorts would be critical[27] was viewed with skepticism even by his otherwise sympathetic biographer.[28] Through the prewar buildup, and after the United States' entry into the war, Arnold's view of the fighter escort question was less than clear and steady. Although the substantial losses of unescorted bombers described in journalistic accounts and intelligence reports from Spain, China, and Britain, as well as heavy losses among the few B-17s used by the RAF in daylight raids in early 1941, seem to have had no effect on the essential linearity of U.S. doctrine, the problem had not fallen wholly out of view. Arnold did reject several proposals to produce large numbers of drop-tanks during that period, but several different Air Corps boards considered heavy fighter development from 1935 to 1941. They recognized the limits of technical development as concern for the safety of the rear of bomber formations increased in light of the British experience. RAF leaders, speaking with the authority of operational experience to the U.S. airmen on the sidelines, expressed doubts about the feasibility of developing escorts, and suggested modifying heavy bombers to serve as escort gunships.[29]

Although a primary, but not sole solution to the fighter escort problem proved to be auxiliary fuel tanks or "drop-tanks," they lay well back in the shadows as the U.S. Army Air Forces shaped their battle plans in the late summer of 1941. Such devices were mentioned in an Air Corps Tactical School text of 1931 and had been requested by Maj. Gen. Frank Andrews when he commanded GHQ Air Force in 1935, but to augment range in point-to-point flying and ferrying, not to aid fighters in battle.[30] The subsequent development efforts fell far short of auxiliary tank programs in the Luftwaffe and Japanese air forces, although tanks were procured for ferrying aircraft from North America to Britain.

The Army Air Corps directly addressed the fighter escort question in the mid-1930s with the development of the Bell YFM-1 AIRACUDA. Air Corps funding had increased enough by the mid-1930s to allow placement of an initial contract with Bell Aircraft to build the only U.S. version of a heavy, multiseat fighter. Designed to offset the speed and firepower of defending single-engine fighter-interceptors by accompanying bomber formations, and fending off such attackers,[31] the XFM-1 (the experimental prototype) and its variants, the YFM-1 and -2, had several special features. The AIRACUDA was the first U.S. aircraft with a "greenhouse" cockpit and electrically powered controls, and its main armament was two 37 mm cannons, served by two of its crew of five who crawled through the wings to nacelles forward of the Allison engines that drove pusher propellers. As with the B-17, the side-blisters mounting .50-caliber machine guns were found to cause too much drag, and sliding panels were substituted. The AIRACUDA's complex electrical systems were powered by auxiliary power units, and the YFM-2 had tricycle landing gear.

In spite of much popularizing of their futuristic image, and being widely seen as a major Air Corps operational type, only thirteen YFMs were built. Although technical problems abounded, the AIRACUDA project enhanced Bell's production capacity, and its extended driveshaft technology was applied to the main Bell types in World War II, the P-39 AIRACOBRA and P-63 KINGCOBRA, and, later, in helicopter production, and in the building of the first American jet aircraft, the P-59 AIRACOMET.[32] Pictures of it were widely featured in print media. Its image as a modernistic superweapon was augmented when the Canadians were denied even "so much as a peep" at it in the summer of 1938.[33] As calls for multiengine fighters appeared in the aviation trade press,[34] General Arnold declared the AIRACUDA the "most striking example" of advanced air technology "anywhere in the world."[35] It was, however, a technical and operational anomaly. No AIRACUDAS saw operational service, and the Air Corps entered the war without a long-range escort fighter. After Pearl Harbor, the thirteen YFM-1s were ordered to Chanute Field, Illinois, and used in training activities. None went overseas; each had cost more than a B-17. More importantly, a crucial doctrinal gap remained open.

Several other nations' air forces went much further in developing heavy escort fighter types in the 1930s. Perhaps the best known of the short-lived type was the Messerschmitt Bf 110, the product of Luftwaffe maneuvers in the winter of 1933–34 that revealed a need for high-altitude reconnaissance and light bomber aircraft, leading to a *Kampfzerstorer,* a heavy fighter that could fly 240 miles per hour within a 600-mile radius. After the Focke-Wulf and Henschel firms produced prototypes, the Luftwaffe technical office, recognizing that development would change specifications, called for a twin-engine, all-weather fighter with three hours' endurance at 350–370 miles per hour.[36] The first prototype Bf 110 that flew in 1936 mounted two 20-mm cannons and five 7.9-mm machine guns and could range 280 miles at 336 mph. A superior design by Kurt Tank, which might have fared much better than the Me 110, the Focke-Wulf 187 twin-engine *Zerstorer* was shelved.[37] After a quarter of the Bf ll0s were lost in the campaigns of 1939-40,[38] the remainder were allocated to antishipping, fighter-bomber, and mainly night-fighter service. Refinements of the type, the Me 209 and 210, and the Heinkel 100, were mainly employed in the latter roles, or in tactical ground support.

The French variant, the Potez 63- heavy fighter series, resembled the Bf 110 both in specifications and in appearance, and was also assigned other roles after suffering heavy losses against single-engine fighters in daylight.[39] The 630 so closely resembled the Bf 110 that many were lost to friendly fire early in the 1940 campaign, leading to the use of the gaudy red and-yellow cowling stripes that distinguished Vichy French aircraft.

The British followed a different path. The Royal Air Force's heavy fighters were not designed to escort bombers, but grew out of the search for a long-range interceptor and a desire to mount heavier armament on fighters. Results were mixed. The durable Bristol BEAUFIGHTER appeared in a series of marks throughout the Second World War, mounting various arrays of machine guns and cannon in the nose, a product of the British concept of "lethal density," that is, a high volume of fire during the brief moments of alignment with enemy aircraft during aerial combat. RAF research pointed up the value of increasing projectile size versus rate-of-fire, but surges in production of existing types as war approached, followed by the demands of combat, deferred the mounting of heavier automatic weapons in fighters or bombers to later in the war.[40] After the Boulton Paul DEFIANT, a large single-engined fighter with a power-driven turret, failed as an attempt to put flexible firepower aboard a fighter, the BEAUFIGHTER, which could carry heavy radar gear, became the principal British night-fighter, and, like other heavy fighter variants, was diverted from daylight air-to-air combat to various auxiliary roles.

The Royal Air Force, daunted by losses and lack of effect in its daylight bombing efforts, shifted to night operations early in 1941, partly because of the short range of its HURRICANE and SPITFIRE fighters. Its single-seat heavy fighter types, the Hawker TYPHOON and TEMPEST, were dogged by combat failures and technical setbacks, and neither played a long-range escort role. Originally designed to mount a dozen .303-caliber machine guns to gain lethal density, the TYPHOON's combat debut was ill-starred. After several of the type's tails came off in pulling out of dives, and the flaw was corrected, it became a close air support mainstay. Its ability to lift heavy arrays of bombs and rockets and its cannons made it the scourge of Wehrmacht ground forces in northwest Europe, 1944–45, and a refinement, the TEMPEST V, became a "hot fighter" and the major killer of German V-1 rockets.[41]

The Soviet approach to the escort fighter problem in the 1930s included development of the Ant-7 "Air Cruiser," the Ant-21 gunship, and experiments with parasite fighters.[42] After specifications for an escort fighter were issued in 1938, Andrei Tupolev's senior designer, Vladimir Pyetlatkov (who, like some other designers and teams, worked in an NKVD research compound in a labor camp) produced a prototype *vyostny perekhvatchik* (high-altitude interceptor) that evolved into the Pe-3 heavy fighter, resembling the Bf 110 and Potez 63- series in appearance, performance, and heavy losses in combat. After Pyetlatkov's death in a plane crash in early 1942, his original concept was revived as the Pe-2V1 *vyostny istrebitel'* (high-altitude fighter) designed to counter German high-altitude raids against Moscow. After that threat faded, the program was canceled in 1944.[43]

The discounting of the heavy losses suffered by unescorted bombers in China in Western defense circles was partly due to the view of many of the Japanese as primitive, some of whom concluded that insurmountable cultural and genetic barriers would prevent Japan's waging air war against Western nations.[44] Much to the contrary, the Japanese army reacted with alacrity to the unexpectedly heavy loss of bombers over Chinese cities and issued specifications for a twin-engine heavy fighter in March 1937. Of three aircraft manufacturers' responses, one evolved into the Kawasaki Ki-45 TORYU (Dragon Killer). Several other types developed during the war included the Ki-95, Ki-96, and Ki-102 and 102a, the latter designed to close the night fighter gap at the end of the war. The Ki-45, like the Bf 110, Potez 63- series, and the Pe 2 and 3, was assigned night-fighter and fighter-bomber roles after suffering heavy losses in daylight operations against enemy single-engine fighters.[45]

In the United States, in the late 1930s, the "convoy" escort fighter aircraft proposed by Billy Mitchell was resurrected by Carl Spaatz, later commander of U.S strategic air forces in Europe in World War II, and first chief of staff of the U.S. Air Force.[46] On the eve of the decimation of the heavy fighter class, it was widely seen as a viable type of combat aircraft, analogous to "pocket battleships."[47] In spite of those heavy losses, the escort fighter remained under scrutiny by U.S. airmen. While drop-tanks came back into focus early in the war, heavier escort planes, along with the phasing of short-range fighter echelons, were seen as a way to provide constant cover for bombers to and from the target area,[48] by several junior Army Air Corps officers.[49] Nevertheless the Air Staff rejected a Wright Field (Materiel Command) proposal for mounting an auxiliary tank system developed at the expense of Republic Aviation on the P-36C. Arnold, now chief of the Army Air Corps, waxed skeptical about the bomber enthusiasts' optimism, and the Materiel Command examined other approaches. One was a modification of the B-10 bomber, and the other, a two-seat pursuit, traded speed for weight and range, embodying the basic dilemma that developers had grappled with in several countries.[50] In spite of those efforts, the escort problem remained on the periphery of concerns, drifting in and out of view, although the American stratagem of relying on repetitive, symmetrical mass arrays and pressing on against stiff defenses had become fully apparent to the Germans.

In 1937, for example, as the AIRACUDA appeared, Alexander P. de Seversky, the aircraft designer-manufacturer who became the best-known civilian proponent of air power in the United States during World War II, advocated a multipurpose, single-engined "convoy fighter."[51] He continued to call for such a type when war came and afterward claimed that Arnold had rejected his 1938 proposal for adding fuel tanks to the Seversky single-engine P-35 fighter.[52] Arnold had not wholly lost sight of the problem and in late 1937 suggested a need for a multiseat, long-range fighter. That led the Air Corps Board to set specifications for a fighter able to fly over 25 percent faster than bombers, carrying one flexibly mounted .30-caliber and four fixed .50-caliber machine guns, with a high rate-of-climb and the ability to carry five or less bombs. Nothing came of it, since resources were still scarce, and the YFM-1 was in hand.[53]

Three years later, in April 1940, the Emmons Board, reviewing American aircraft construction priorities in view of the course of the air war in Europe, urged moving fighters down and escort fighters up on the priority list.[54] American interest in the problem outlived that of the British, but the heavy losses of French and German heavy fighter types and the Royal Air Force arguments about limited fuel capacity brought American airmen closer to their skepticism. The shapers of the air war master plan AWPD-1 ranked development of such a type last in their list of production priorities,[55] and the auxiliary fuel tank fell once again back into the shadows. As American entry into a European air war loomed larger in mid-1941, three months before Pearl Harbor, in a coauthored article in *Flying,* Arnold and then Major General Ira Eaker noted the failure of multiplace fighters in Europe to accompany "bombers into enemy territory to ward off enemy pursuits" since they tended to be "too slow and not sufficiently maneuverable." Ironically, the article was illustrated by a photo of a YMF-1, upon which so much praise had been heaped not long before.[56] A year later, when masses of RAF fighters in ratios between 7:1 and 10:1 escorted U.S. bombers on their ever-larger daylight missions against French ports, that formidable array was somehow not related to their low losses. That may have been because of the distracting problems with logistics, administration, and maintenance that beset Eaker's VIII Bomber Command, the core of 8th Air Force. In any case, the crucial doctrinal anomaly was left unaddressed as the grinding continued.

Earlier in 1942, the renewed Nazi offensive in Russia had led the RAF to make massive fighter sweeps to take the pressure off the beleaguered Soviets. The early U.S. bombing raids, although marginal in impact, also helped to stir up German fighters, although their losses were usually far less than Allied estimates.[57] A pessimistic review of British bombing efforts was under way as a public clamor for a "Second Front" rose, first in Britain and then in America, echoing Soviet calls heard since August 1941 for a major landing in northwest Europe to divert the Nazis. Those calls rose to a peak at the bleakest time in the war for the Allies, the summer of 1942, but things had brightened considerably when the Casablanca Conference convened in early 1943. There, the RAF supported Eaker's plea to Roosevelt and British prime minister Winston Churchill that U.S. bombers not be merged with RAF Bomber Command's night raids, but that they mount a separate daylight precision bombing effort, as part of a "round-the-clock bombing" campaign. The president and the prime minister approved the proposal for the Combined Bomber Offensive, later known as Operation POINTBLANK.

The low losses of U.S. bomber forces operating with heavy fighter cover over France in late 1942 and in a raid against Wilhelmshaven in January 1943 fueled Eaker's optimism and made believers out of doubters. In 8th Air Force, optimism tended to occlude the fact that various fighter range-extension projects under way bore no immediate promise of having an effect on operations. Calls were mounting from various quarters throughout the USAAF for help in that regard, and board after board and conference after conference added impetus to that concern even before the heat of battle added a new urgency.[58] Paradoxically, the strategic situation had appeared to ease somewhat by early 1943, as drop-tank purchasing began in earnest. Enthusiasm for the B-17 FLYING FORTRESS's ability to survive daylight missions had infected

American air leaders, including Spaatz. When serving on the Air Staff in Washington in May 1942, he thought U.S. bombers might join the RAF in night bombing if fighter escorts could not accompany them against targets in Germany.[59] In August, Spaatz accepted Arnold's and Eaker's rosier view, and the myth of a German "fighter belt" that could be fought through, even though Allied fighter pilots warned that longer missions would exhaust bombers' ammunition quickly, and that there was no such "belt."[60]

Eagerness subsided momentarily on February 26, 1943, when 7 of 63 bombers were lost against Wilhelmshaven, but then rose again when only 2 of 97 aircraft were downed over Vegesack on March 18. All bombers returned from attacks on Ford and GM plants at Antwerp on May 4, with P-47s assisting. Then, in mid-May, the storm broke in full fury, and raged until October. Loss rates sometimes surged above 10 percent per mission, and no major raid into Germany was escorted by fighters until late September. On October 18th, General Anderson, head of VIII Bomber Command, told his commanders that "we can afford to come up only when we have our fighters with us,"[61] an admission that marked the end of a vision of air power bringing victory without an invasion of "Fortress Europe" by ground forces.

How had that larger doctrinal failure come to pass? Although denied the status of an independent service, the USAAF had gained substantial autonomy even before Pearl Harbor through a quasi-diplomatic alliance with the RAF, its counterpart—and separate—service. Although substantial USAAF resources were allocated to directly support army ground operations, doctrine and practice varied among the various Theaters of Operations. U.S. airmen, given a relatively free hand in planning their strategic bombing operations against Germany, had recognized a need for escort fighters in AWPD-1, drafted in August 1941, but no such concern appeared in the successor plan, AWPD-42.[62] By that time, as noted earlier, in keeping with the aphorism that the Devil is in the details, American airmen encountered a host of unforeseen problems, many of which remained unsolved as the 8th Air Force's bombers began to operate beyond escort fighter range in the spring of 1943.[63] Many stopgaps were employed. By midyear, with an average per mission of almost 20 percent of bombers being hit by fighters, a crash body armor program was launched that reduced wounds by 16 percent. Mounting losses led to a reduction of 8th Air Force bomber crewmen's required missions from fifty to twenty-five, but even with the higher loss rate, that yielded only increased survival chances from about eight to twenty-four percent.

In 1943, in stages, the 15th Air Force opened a new air war front, flying first from bases in North Africa and then in Sicily and Italy, operating sometimes alone and sometimes with the 8th, but casualty rates climbed in both commands. From August to October, beyond the 8th's average losses of nearly 15 percent on its missions, many returning planes were damaged beyond repair, and many of their crewmen killed and wounded. With repair facilities swamped, and the weather closing in, bombing praxis eroded doctrine as formations dropped bombs en masse, following the lead of a master bombardier, with enlisted men acting as "toggleiers."

As aborted missions and turnbacks increased, full realization of the state of affairs came slowly at the top; in late May 1943, Eaker admitted at a press conference that new German tactics and devices were "damaging," but predicted operations would "continue . . . at a loss ratio of five percent or less." A month later, after the 8th suffered 14 percent losses on one mission and 9 percent on another, Eaker cabled Arnold for more replacement crews, depot facilities, and drop-tanks. Arnold bent every effort to respond, but was unable to close the drop-tank production gap which widened in 1943, in a modern version of "for want of a nail," despite attempts to close it with increasing desperation.[64] As has been seen, auxiliary tanks were not a conceptual bolt from the blue. Both the Japanese and German air forces had used auxiliary fuel tanks in large numbers,[65] and the USAAF had employed range extension tanks to ferry some P-38s and P-47s to Britain from late 1941 on.[66] The relative trickle from the United States led to contracts being placed in Britain after increasing Nazi U-boat sinkings of Allied merchant ships in 1942–43 stanched the small flow even further. Large orders were also placed in the United States in January 1943, but the mismeshing of logistical and operational subelements of the USAAF, Materiel Command snarls, and delays from British sources all combined to crimp that logistical pipeline until December 1943. To further complicate matters, in the first long-range escort missions, the tanks originally designed for ferrying proved less than adequate in the stresses of combat flying.[67] From the standpoint of complexity theory, it was a demonstration of small details yielding large outcomes.

Rising concern over the drop-tank delay at high echelons and in operational units was matched by rising public awareness of the escort fighter gap. Some, like William Bradford Huie, claimed the B-17 was really a "Flying Fortress," while others argued that only twin-engined heavy fighters could provide adequate escort, pointed to the need for drop tanks, or claimed that the Allies had an "abundance of long-range fighter escorts."[68] With its doctrine crumbling on the anvil of combat, the 8th launched a major effort to improve gunnery through analysis and training, while Curtis LeMay and others developed a tiered "combat box" formations to present German fighters, who usually closed head-on, with a vastly increased mass of firepower. The most bizarre stopgap appeared in the form of the YB escort bomber concept, a response from a board headed by World War I ace Brig. Gen. A. J. Lyon that called in August 1942 for "flying battleships." These, like the Messerschmitt 110, were seen as analogous to a naval type and labeled "destroyer escort planes"; a few B-17s and B-24s were converted into YB-40s and YB-41s, respectively. Tracks to convey masses of ammunition to the waist and tail gun position to seven pairs of .50-caliber guns, extra armor and clear plastic noses were installed and, on some, extra dorsal power turrets. YBs carried no bombs, but some 40,000 rounds of ammunition.[69] Providing a only a very localized augmentation of firepower, the YBs flew as fast or faster than bombers approaching targets, but slower than bombers that had dropped their loads. The crews of the bombers were not keen to hang back with their lumbering companions. The YB program dead-ended in mid-1943, as did a fitful attempt to convert Martin B-26 MARAUDERS to an escort version.[70]

Each side constantly modified tactics and equipment to counter their opponents' stratagems. The Luftwaffe fighters abandoned head-on attacks and began hanging back and above, using standoff cannons and rockets, against which there was no defense except escort fighters. The myth of a "fighter belt" proved to have been wishful thinking. Diverting their elite night-fighters to daylight operations, the Germans used long-range rockets and heavier aircraft cannon, and dropped bombs onto mass formations. That allowed them to hammer the new box formations while remaining out of reach of their massed firepower. While innovations in tactics and bigger antiaircraft guns paralleled those developments, the operational research section of 8th Air Force estimated that German fighters alone caused 40 percent, and were the final cause of 70 percent of losses. Even when they did no direct damage, the standoff attacks by heavy fighters during bombers' final approach to targets disrupted their formations, confounding attempts at pinpoint bombing.

All came right for the USAAF daylight strategic effort in the end with the coming of masses of escort fighters, but not all at once, nor did the tide of battle turn sharply. When the long-awaited drop-tanks arrived in large numbers, after an agonizingly slow progress, they proved less than a full solution to the range extension problem. Not a simple technology, drop-tanks had not been popular in the Luftwaffe, since they hindered maneuverability and were a fire hazard, especially in aborted takeoffs. They required a complex pumping system to function at high altitudes, along with linkage and mounting mechanisms that required several dozen parts. Fortunately, experience with ferrying tanks, although they proved unsuitable, helped in engineering and producing types that could stand up to the demands of aerial battle. There were other problems, however. Because their drag cut the small amount of extra speed that gave Allied fighters a "performance edge" in combat, drop-tanks were often released when Luftwaffe fighters were sighted. When German airmen took advantage of that by feinting to trigger drop, Allied fighter pilots countered by draining fuel from their drop-tanks first.

At the same time, production of main fighter aircraft types was modified to provide additional on-board tankage. An escort variant of the twin-engine Lockheed P-38 LIGHTNING, originally designed as a long-range interceptor, was joined in growing numbers by a heavy single-engine fighter with greater internal fuel capacity, the Republic P-47 THUNDERBOLT, the first type of which had flown in May 1941, and a lineal descendant of the P-35. Later, versions of the P-51 MUSTANG filled the gap in stages, at the same time that a series of tentative long-range fighter projects proved fruitless.

The first major deliveries of drop-tanks finally came as severe winter weather curtailed operations, in the aftermath of a major command shake-up. Carl Spaatz assumed command of the U.S. Strategic Air Forces in Europe in the summer of 1943, and in the autumn, Eaker, much against his wishes, was sent to command Allied air forces in the Mediterranean, Jimmy Doolittle taking his place. The latter revised tactics, using the bombers as bait to force the Luftwaffe fighters to confront the growing numbers of U.S. long-range fighters. Most daylight bombing became limited area attacks, and efforts were focused in stages on marshaling yards, industrial plant complexes, and synthetic oil facilities and refineries.

The aerial "Battle of Germany," which extended over the next year, opened with a series of major raids, from "Big Week" in mid-January 1944 on through to mid-April. Overall, just over 400 aircraft were lost out of some 8,200 sorties, a decline from just over 10 percent in the first three major missions to about 2 percent at the end. As those figures suggest, the Luftwaffe did not abandon the field all at once. Postwar examinations of Nazi records reveal how much Allied intelligence overestimated Luftwaffe fighter losses, and how German fighter production rose from Albert Speer's reorganization and dispersal of the German industrial system in 1943. Although bombing pressure on the Reich had eased momentarily in the spring of 1944 as Allied air forces launched massive attacks in preparation for Operation OVERLORD (D-Day), the pace of battle quickened again as daylight strategic bombing aimed at Nazi oil production. That decision offset the increased German piston-engined fighter production and the ability of the jets that appeared in late 1944 to operate in effective numbers. The expanding hordes of Allied fighters (fourteen MUSTANG groups accompanied one raid in mid-March 1945) overwhelmed German defenses. The pressure increased after the Allied ground forces broke out of Normandy in August 1944 and fighter bases moved eastward behind the advancing

Allied armies. With the air battle of Germany won by early 1945, major bomber raids against the Reich's cities were suspended, and the Allied tactical air forces roamed German airspace with near impunity.

Why was a flawed sword wielded well after it showed signs of shearing? Was it because the U.S. daylight bomber offensive was planned and led by officers with little operational experience? Although doctrines, training, planning, and procurement remained out of phase with operational realities for two years after Pearl Harbor, U.S. airmen did readily abandon other peacetime doctrines, such as attack aviation and dive-bombing, when those were found wanting. Were the daylight efforts maintained for political reasons, because they represented a central rationale for a separate air service, or because Americans did not want to be seen as having joined the RAF in "throwing strategic bombing at the man in the street"? Did the "bomber barons" really believe that they had a war-winning weapon? Was the optimism of mid-1943 the product of "group-think" at VIII Bomber Command, or Eaker's dynamic leadership, or, indirectly, of the burning zeal of "Bomber" Harris, chief of RAF Bomber Command? Considering those uncertainties from the perspective of complexity leads to another question: Can such causal tendrils really be sorted out to produce anything beyond an ambiguous conclusion? Since tracing events and searching for patterns and causes is based on extant documents, there is no way to measure such abstruse factors as the stylistic self-restraint of air crews in debriefing, the filtering and sanitizing in processing such "feedback," and chronic overestimating of operational impact. Debates over targeting policy and the apparent lack of creativity in wielding the air arm in 1943 are likely to continue, as they are in respect to other air campaigns.

While hindsight has tended to make such judgments increasingly harsh, considering such events with complexity and chaos in view offers some sense of how much both doctrine formulation and operational application are laced with intractable "fuzziness." Thus, these three cases, selected to illustrate the crumbling of doctrine in the course of operations, are only impressionistic representations of broad and complex processes. The intricate and contradictory interplay of many forces and perceptions seen in each can also be found in other major military undertakings. In both World Wars, substantial potential advantage was dribbled away in the initial employment of various weapons, most notably U-boats, poison gas, "Q-ships," and tanks. In all those cases, major commanders were trying to monitor a tangle of complexities and close the gap between their perceptions and visions and a not fully perceived—or perceivable—reality. There is also some similarity in the apparent failure to think out alternatives about what systems might be used for if doctrine and expectation did not mesh with actual events, and it turned out that, as Admiral Beatty observed at Jutland, something was very, very "wrong." Since the exact dynamic of how things went awry cannot be portrayed to full exactitude, it is not as easy to judge what should have been done as historians often tend to do.

It should also be noted that there are several basic differences between the various cases. The momentum of the battle cruisers and daylight massed bomber formations was maintained in the face of diminishing returns, but that was not the pattern with the tank destroyers. While none of the three systems was really used as intended initially, in the case of the battle cruisers, the concept was the "baby" of a powerful personality, and it glided a long way on momentum once initiated. In the second case, daylight strategic bombing doctrine was both formulated and kept under way by dedicated individuals, but in the third, the tank destroyers lacked a visible champion and proverbial "friends at court." One cannot measure the interplay and collective effect of such varied forces as technological evolution, production snarls, competing adaptation and tactics, doctrine, and such "fuzzy" human dimensions as politics, bureaucratic process, and the interaction of personalities, but the influence of some slight variations on the flow of action is obvious enough, including individual human behavior. Fisher and Eaker were tough, shrewd, and persuasive and possessed strong "command presence"; their struggles to transmute their visions into reality suggests both the power of individual tenacity in the complex dynamics of warfare, and how the "fog of war" can flow down the chain of command as well as upward.

Such hierarchical gravity in military systems, and other complex organizations as well, affirms the adage that the "whim of the prince has . . . the power of law," and affects the formulating and applying of doctrine when leaders strike postures of surety and aggressive determination, irrespective of any objective reference. Again, the special irony here is that such role-playing is widely seen, and not only in military circles, as a virtue among leaders, to the extent that those who have failed have been admired for tenacity in the face of certain doom, from Horatius at the bridge, and Roland at Roncesvalles, to Bonaparte, Robert E. Lee, and Rommel. The resultant tendency to focus on major leaders' roles and style tends to blur the far more complex dimension of the behavior of followers, especially key functionaries. The influence of likes, dislikes, intrigue, and favoritism give both flavor and substance to military history as it does to other human activities. There are many cases of "sharp elbows," that is, rivalry and factionalism (known as the "war within the war" in the German army in World War I). Such instances as Sir Douglas Haig's displacing of Lord French in 1915 and Pershing's "exiling" of MacArthur are not easy to track in official documents, nor are the full depths of personal motives. Both frictions and affections may be wholly out of view, or barely glimpsed. Even as a vague element, such informal and personal processes add a layer of critical complexity onto the others. However immeasurable, they constitute slight influences that, if visible to historians at all, have effects out of proportion to apparent significance, like the personal animus between Russian commanders Samsonov and Rennenkampf in 1914.

At a more formal and visible level, the mechanisms of deference and forced compliance in military institutions have often suffused doctrine and command decisions with an aura of optimality, and sometimes infallibility, investing high authority with exaltation and giving a premium to the imposing of order without concern for its function. To the extent that both doctrines and leadership arise from such urges to impose order on confusion, these cases suggest other quandaries. Military historians can ponder whether sharply defined doctrines, or those that were loosely shaped, realigned hastily, or based on the acceptance of the likelihood of uncertain operational contexts have been more successful. Military professionals will be more concerned as to whether doctrine can be useful if it is more fuzzy than crystalline, and accepts the unexpected and the chaotic to the point of anticipating the possible failure of the doctrine itself. It could be argued that a military organization, especially in peacetime, the politics of funding and procurement aside, cannot function without an orienting axis in the form of doctrine. Notably, various branches and services have had doctrines that clashed with others', formally and informally, such as the U.S. Navy's "Maritime Strategy" of the 1980s and the U.S. Army Air Corps Tactical School's shadow doctrine in the 1920s and '30s. A further irony lies past that dissonance: the fact that the success of doctrine in one context, such as the Nazi blitzkrieg, the Italians' underwater warfare and light naval units, and the Japanese kamikazes, may yield substantial tactical success but not overall victory.

Since apparently minor, informal, or minuscule forces may nudge process toward chaos as a component in an elaborate variable equation, what can be seen may not even closely approximate what is, either by design, propaganda or camouflage, or accident—or because what *is* may be beyond the limits of perception, individual or systemic. As Ralph Strauch has pointed out, "Perception and interaction are inexorably intertwined and cannot be separated."[71] Nevertheless, not only military professionals and historians, but most conscious actors in Western culture, are habituated to seeing their worlds of thought and action as separate compartments set in hierarchical tiers arrayed in terms of rationalistic, linear concepts and categories. Given the strength of the reflex implanted in military professionals (but others as well) to seek order and simplicity in complex organizational dynamics or warfare and to dominate, it seems unlikely that those thus acclimated would welcome the idea that understanding might be gained by accepting the possibility that there are limits to understanding. A salient implication is that because determining exactly where those, and even more general, outer boundaries lie is rarely possible, choice and action lie closer to the analogies of a performing art or gambling than to a sequenced, rational process. Yet some signs of a sea change have become visible over the last generation, perhaps a portent of the rise to power of those who have grown up in a milieu of computers and systems sensitivity; as a result, this new generation may more easily recognize the irreducible ambiguities that affect military doctrine if it is tested in the crucible of war.[72]

NOTES

1. Experts' imprecision can be found outside the military realm, for example, crowd-size estimation, weather forecasting, and the nuclear-thermonuclear bomb tests (all the bomb tests exceeded advanced estimates of yield by scientists by about 300 percent).

2. For operational statistics, see Roger A. Beaumont, *Military Elites* (Indianapolis: Bobbs-Merrill, 1974), pp. 158–60 and 205.

3. W.S. Chalmers, *The Life and Death of David, Earl Beatty* (London: Hodder & Stoughton, 1951), p. 120; John Jellicoe (Admiral Viscount), *The Grand Fleet 1914–1916: Its Creation, Development and Work* (New York: George H. Doran, 1919), pp. 304–306; and Arthur J. Marder, *The Anatomy of British Sea Power: A History of British Naval Policy in the Dreadnought Era, 1880–1905* (Hamden: Archon, 1964), pp. 103 and 334.

4. Peter Smith, *British Battlecruisers* (New Malden, Surrey: Almanack Press, 1962), p. 7.

5. Richard Hough, *Dreadnought* (New York: Macmillan and Co., 1964), pp. 86 and 102.

6. Ruddock F. Mackay, *Fisher of Kilverstone* (Oxford: Clarendon, 1937), pp. 270, 326, 389, and 426; also see R.H. Bacon [Admiral Sir], *The Life of Lord Fisher of Kilverstone, Admiral of the Fleet,* Vol. 1 (New York: Garden City Publishing Co., 1929), pp. 215 and 250.

7. Jon Sumida, *In Defense of Naval Supremacy* (Boston: Unwin Hyman, 1989), p. 258.

8. Norman Friedman, *U.S. Battleships: An Illustrated Design History* (Annapolis: U.S. Naval Institute Press, 1985), p. 144.

9. Mackay, *Fisher of Kilverstone,* pp. 443, 468–69, and 473–74; and Raven Law and John Roberts, *British Battleships of World War Two* (Annapolis: U.S. Naval Institute Proceedings, 1976), p. 45.

10. James Goldrick, Introduction, in Filson Young, *With the Battlecruisers* (Annapolis: U.S. Naval Institute Proceedings, 1921 [repr. 1986]), pp. xvi–xvii; Arthur J. Marder, *From the Dreadnought to Scapa Flow: The Royal Navy in the Fisher Era, 1904–1919,* Vol. 3, *Jutland and After [May 1916–December 1916]* (London: Oxford University Press, 1960), p. 174; John A. Roberts, *Invincible Class* (Greenwich: Conway Maritime Press, 1972), pp. 45–47; and N.J.M. Campbell, *Jutland:*

An Analysis of the Fighting (Annapolis: U.S. Naval Institute Press, 1986), pp. 371–374.

11. R.H. Bacon, *The Life of John Rushworth the Earl Jellicoe* (London: Cassell, 1936), p. 319.

12. Ernie Bradford, *The Mighty Hood* (Cleveland: World, 1959), p. 31.

13. Main sources for this section are: *Tank Destroyer Units*, Doc. N-12472-17, U. S. Army Command and General Staff College Library; Emory A. Dunham, *The Tank Destroyer History* (Washington, D.C.: Historical section, Army Ground Forces, 1946); Ralph Lang, *Tank Destroyers*, [Monograph for Advanced Officer Class No. 1] (Fort Knox, Ky.: The Armor School, 1947; H.T. Mayberry, "Tank Destroyer Battle Experience," *Military Review*, 25:5 (May 1945), pp. 23–27; Charles M. Baily, *Faint Praise: American Tanks and Tank Destroyers During World War II* (Hamden: Archon Books, 1983); and Roger Beaumont, "'Seek, Strike, Destroy'—The World War II Tank Destroyer Corps," *Armor*, 80:2 (March–April 1971), pp. 42–46.

14. See "The Tank Killers," *Fortune*, 26:5 (November 1942) , pp. 116–118 and 181–183; and Hubert Thorner, "The Tank Destroyers and Their Use," *Military Review*, 23:1 (January 1943), pp. 21–24.

15. Richard M. Ogorkiewicz, *Armoured Forces: A History of Armoured Forces and Their Vehicles* (New York: Arco Publishing Co., 1970), p. 201.

16. James Parton, *'Air Force Spoken Here': General Ira Eaker and the Command of the Air* (Bethesda: Adler & Adler, 1986), pp. 280–281.

17. Herman S. Wolk, *Planning and Organizing the Postwar Air Force* (Washington, D.C.: Office of Air Force History, 1984), p. 17.

18. Bernard Boylan, "The Search for a Long-Range Escort Plane, 1919–1941," *Military Affairs*, 30:2 (Summer 1966), pp. 58–59.

19. See Enzo Angelucci & Peter Bowers, *The American Fighter* (New York: Orion Books, 1988), pp. 39–41, 97–99, 99–101, 177, 271–272, 296–297, and 367.

20. Giulio Douhet, *The Command of the Air*, trans. Dino Ferrari (New York: Coward McCann, 1942), pp. 346, 359, 376, 377, and 380.

21. Thomas A. Fabyanic, *Strategic Air Attack in the United States Air Forces: A Case Study* [Air War College/Air University Report No. 5899] (Manhattan, Kans.: Military Affairs/Aerospace Historian, 1976), pp. 26 and 31.

22. Bernard Boylan, *Development of the Long-Range Fighter Escort* [U.S. Air Force Historical Studies No. 136] (Maxwell Air Force Base: Research Studies Institute, U.S. Air Force Historical Division, 1955), pp. 12–13.

23. Wanda Cornelius & Thayne Shore, *Ding Hao: America's Air War in China, 1937–1945* (Gretna: Pelican Publishing Co., 1980), p. 48; for a critique, see Heywood S. Hansell, Jr., *The Air Plan That Defeated Hitler* (Atlanta: priv. publ., 1972), p. 22.

24. Claire Chennault, "The Role of Defensive Pursuit," in *Coast Artillery Journal*, Pt. 1, 76:6 (November–December 1933), pp. 412–416; Pt. 2, 77:1 (January–February, 1934), pp. 5–11; and Pt. 3, 77:3 (March–April 1933), pp. 87–93.

25. DeWitt S. Copp, *A Few Great Captains* (Garden City: Doubleday, 1980), pp. 257–261.

26. Fabyanic, *Strategic Air Attack . . .*, p. 31.

27. Henry H. Arnold, *Global Mission* (New York: Harper and Brothers, 1949), pp. 149 & 376.

28. Thomas M. Coffey, *Hap: Military Aviator* (New York: Viking Press, 1982), pp. 167–168.

29. William R. Emerson, "Operation POINTBLANK: A Tale of Bombers and Fighters," in Harry R. Borowski, ed., *The Harmon Memorial Lectures in Military History, 1959–1987* (Washington, D.C.: Office of Air Force History, 1988), pp. 451–452.

30. Boylan, *Development of Long-Range Fighter Escort*, pp. 16–17.

31. I.B. Holley, *Buying Aircraft: Materiel Procurement for the Army Air Forces* (Washington, D.C.: Office of the Chief of Military History, 1964), pp. 200–201; Ray Wagner, *American Combat Planes* (Garden City: Random House, 1982), pp. 248–249; and "Airacuda: A Formidable Multiplace Fighter," *Scientific American*, 157:5 (November 1937), p. 307.

32. Donald J. Norton, *Larry: A Biography of Lawrence D. Bell* (Chicago: Nelson-Hall, 1981), p. 68.

33. Holley, *Buying Aircraft*, p. 199.

34. Horace J. Alter, "The Fighting Airplane: Will Multi-Place Ships Challenge the Fighter Forces?," *Ordnance*, 19:109 (September–October, 1938), pp. 79–84.

35. *An Aviation Story: The History of Bell Aircraft Corporation* (Buffalo: Bell Aircraft Corporation, 1945), p. 32.

36. Matthew Cooper, *The German Air Force, 1933–1945: An Anatomy of Failure* (London: Jane's, 1981), pp. 47 and 53.

37. James J. Halley, *The Role of the Fighter in Air Warfare*, ed. Charles W. Cain (Garden City: Doubleday & Co., 1971), pp. 148–160.

38. Charles Cain, ed., *Aircraft in Profile*, Vol. 9 (Garden City: Doubleday & Co., 1971), pp. 148–160.

39. Paul Camelio & Christopher Shores, *Armée de l'Air: A Pictorial History of the French Air Force, 1937–1945* (Warren, Mich.: Squadron/Signal Publications, 1976).

40. Malcolm Smith, *British Fighters Since 1912* (London: Putnam, 1979), p. 268.

41. William Green, *Famous Fighters of the Second World War* (New York: Hanover House, 1958), pp. 99-105.

42. Alexander Boyd, *The Soviet Air Force Since 1918* (New York: Stein & Day, 1977), p. 62.

43. William Green & Gordon Swanborough, *Soviet Air Fighters*, Pt. 2 (New York: Arco Publishing, 1978), pp. 3–5; also see Cain, *Aircraft in Profile*, Vol. 10, p. 100; and Enzo Angelucci, *The Rand McNally Encyclopedia of Military Aircraft, 1914–1980* (New York: The Military Press, 1983), pp. 293–294.

44. E.g., Leonard Engel, "Japan Is Not an Air Power," *Flying and Popular Aviation*, 28:1 (January 1941), p. 72.

45. Rene Francillon, *Japanese Aircraft of the Pacific War* (London: Putnam, 1970), pp. 95–99, 127, 134, 141–142, and 244.

46. I.B. Holley, "General Carl Spaatz and the Art of Command," in Wayne Thompson, ed., *Air Leadership: Proceedings of a Conference at Bolling Air Force Base, April 13–14, 1984* (Washington, D.C.: Office of Air Force History, 1986), p. 31.

47. Leonard Engel, "Pocket Battleships of the Air," *Popular Aviation*, 36:4 (April 1940), pp. 10–11.

48. Boylan, *Development of the Long-Range Fighter Escort*, pp. 7–10.

49. *Ibid.*, pp. 30 and 46–47, 52.

50. Wesley Frank Craven & James Lea Cate, *The Army Air Forces in World War II*, Vol. 6 (Chicago: University of Chicago Press, 1955), p. 241.

51. Alexander P. de Seversky, "My Thoughts on War," *Popular Aviation*, 36:4 (April 1940), pp. 18 ff.

52. Alexander P. de Seversky, *Air Power: Key to Survival* (New York: Simon & Schuster, 1950), pp. 210–213.

53. Boylan, *Development of the Long-Range Escort Fighter*, pp. 17–18.

54. Holley, *Buying Aircraft*, p. 33.

55. Boylan, *Development of the Long-Range Escort Fighter*, p. 38.

56. Henry H. Arnold & Ira C. Eaker, "Winged Warfare," *Flying and Popular Aviation*, 34:3 (September 1941), pp. 222–223; a photograph of the *Airacuda* appeared in the book *Winged Warfare* (New York: Harper & Brothers, 1941), facing p. 14.

57. F.H. Hinsley, E.E. Thomas, C.F.G. Ransom & R.C. Knight, *British Intelligence in the Second World War: Its Influence on Strategy and Operations*, Vol. 2 (London: Her Majesty's Stationery Office, 1981), p. 270.

58. Boylan, *Development of the Long-Range Escort Fighter*, pp. 113–114.

59. Fabyanic, *Strategic Air Attack in the USAF*, pp. 69–70. Some U.S. units flew night missions with the RAF.

60. Boylan, *Development of the Long-Distance Fighter*, p. 63.

61. *Ibid.*, pp. 101–102.

62. See James C. Gaston, *Planning the American War: Four Men and Nine Days in 1941: An Inside Narrative* (Washington, D.C.: National Defense University, 1982), p. 106; also see Thomas Fabyanic, *Strategic Air Attack in the USAF*, p. 65; and Hansell, *Air Plan That Defeated Hitler*, p. 253.

63. Craven & Cate, *Army Air Forces in World War II*, vol. 6, 1955, pp. 334–335.

64. Arnold, *Global Mission*, pp. 375–376.

65. Harold B. Hinton, *Air Victory: The Men and the Machines* (New York: Harper & Brothers, 1948), p. 267; and Holgar Hanroth, *Die Deutsche Luftwaffe vom Nordkap bis Tobruk, 1939–1945* (Freiburg: Podzum-Pallas Verlag, [1981]), pp. 89, 93, and 96.

66. Craven & Cate, *Army Air Forces in World War II*, Vol. 2, 1949, p. 634.

67. Boylan, *Development of the Long-Range Fighter Escort*, pp. 118–134.

68. E.g., see Hal Goodwin, *Aerial Warfare: The Story of the Aeroplane as a Weapon* (New York: New House Library, 1943), pp. 141–142; Francis Vivian Drake, *Vertical Warfare* (Garden City: Doubleday Doran, 1943), p. 22; and William Bradford Huie, *The Fight for Air Power* (New York: L.B. Fischer, 1942), p. 263.

69. For details, see Lloyd S. Jones, *U.S. Bombers* (Fallbrook, Calif.: Aero Publishers, 1974), pp. 138–143.

70. Boylan, *Development of the Long-Range Fighter Escort*, pp. 136–146.

71. Ralph Strauch, *The Reality Illusion: How We Create the World We Experience* (Wheaton, Ill.: Quest Editions, 1983), p. 182.

72. E.g., V.V. Druzhinin & D.S. Kontorov, *Decision Making and Automation: Concept, Algorithm, Decision* (Washington, D.C.: U.S. Government Printing Office, n.d. [c. 1978]), and the U.S. Army's field manual *FM 100-5 Operations* (Washington, D.C.: Department of the Army, 1993); also see Paul E. Funk, "Battle Space: A Commander's Tool on the Future Battlefield," *Military Review*, 73:12 (December 1993), pp. 36–47.

5

The Chaotic Sense of War

When Marshal Konev observed that "the total picture of the [Second World War] can be formed only of many reminiscences," he also recognized that a full assemblage of such details could not be encompassed within the most exhaustive history of the conflict.[1] That always-lopsided ratio of facts to the limits of the format in portraying the past renders history unavoidably impressionistic. As a result, historians have tended to use personal accounts of combatants more as a veneer, tincture, or spice than as a major structural member in the construction of military history, and leave the latter approach to journalists. Part of that stems from concern that such evidence is unverifiable and may be little more than self-justification, anecdotes, fabrication, exaggeration, or flawed memory. Leo Tolstoi, highly skeptical of military historians' attempts to frame a "big picture," confessed his own inability to grasp the formats of great battles and campaigns, and judged it more important "to me to know in what manner and under the impetus of what feeling one soldier has killed another, than to learn the disposition of troops at Austerlitz or Borodino."[2]

In the twentieth century, then, academic and official historians have tended to deal with war from a broader operational perspective, leaving the "microdynamics" of war mainly to novelists, filmmakers, journalists and memoirists. In the mid-1970s, however, a countercurrent rose to the New Military History, which had eschewed "drum-and-trumpet" combat history. Some military historians wove eyewitness accounts into a tapestry, somewhat in the spirit of such battle dynamicists as Ardant du Picq. Salient examples of that were John Ellis's and Richard Holmes's assemblages of vignettes, which often conveyed a sense of combat-as-chaos. Scattered like ore throughout such works and along the fringes of military history,[3] that corpus remains an uncertain basis of evidence for several reasons, most importantly because of the blurring of perception and memory in the special strains of war. A sense of unreality and detachment has been expressed in many accounts of combat, stemming from the grinding stress, draining fatigue, and indescribable noise that impairs the ability to think and prevents the use of speech; nauseating, numbing fear; and deep horror.[4] In a letter of 1943, the best-known American war correspondent of World War II, Ernie Pyle, complained that "the war gets so complicated and confused in my mind . . . ,"[5] in keeping with many combat veterans' description of battle as fragmented and formless, and what John Ellis deemed "fleeting impressions with no coherent shape."[6] Similar confusion and uncertainty are expressed in the letters and memoirs of literate enlisted men from the mid-eighteenth century onward, as well as those of officers.[7] A Wehrmacht general's report that "the chaos that reigned behind our front" in August 1944[8] matched the sense of blind turmoil conveyed in the works of novelists like Richard Aldington, Jaroslev Hasek, and Fred Majdalany. The U.S. Army tried to offset that effect by producing a set of short histories during and just after World War II for veterans in hospitals. In spite of that, a greater awareness of causes or of what was happening at the time sometimes only heightened a sense of irony. Beyond surreal fictional treatments like Heller's *Catch* 22 and Thomas Heggen's *Mister Roberts,* many observers of warfare, like Tolstoi, judged it impossible to describe coherently or precisely their experience of combat as "fast, unfair, cruel and dirty."[9]

It is not surprising, then, that many veterans felt that enduring the special horror of battle separated them from all others. A veteran of "hard pounding" tank-to-tank fighting in North Africa observed, "There is no easy way to prepare soldiers for this blood and guts of combat. Survivors learn the hard way. A soldier has to learn to kill without emotion and function despite the horrors of war he might witness."[10] Because such stresses have rendered so many combat veterans silent or inarticulate, much of the description of war has been left to others. Some of that has been because of the literal blotting out, by stress and fear, of memory or the warping of it. Thus, the often-heard comment to the effect that "Uncle Bill was in the war, but he never talks about it, even if you ask him." The fear of long-suppressed or denied images being brought to the level of consciousness has led many veterans to avoid discussion of anything even tangential to them, let alone their rendering a detailed account. That paradoxically abandons the ground in a historiographical sense to the proverbial false claims and bravado of nonveterans, "rear-echelon commandos," and "REMs," since many prominent military historians and analysts have had no military service or war experience. When veterans suggest that those who did not share their specific ordeal, even those who underwent other ordeals, cannot comprehend or validly describe what really happened, they may be right, although they themselves may not be able to do so, either. Beyond that lies the irony that very few participants in modern war had control over where they went, and what they did and experienced.

Since documents are seen by professional historians as the primary form of evidence, the unevenness and thinness of such residue of combat is compounded by the scarcity of experiential accounts. That is not merely an academic problem, for the view of warfare of very many people, from adolescents to ruling elites, is therefore derived from the work of journalists, historians, novelists, and filmmakers. That helps explain the difference between the chaos depicted in firsthand accounts of veterans and witnesses, and the relatively neat sequencing and order in histories based mainly on official documents, in commanders' memoirs, and in many fictional and popular images of war. Does that reflect a general pattern of maturation, in which only the passing of time and accumulating of data allow historians and memoirists with a "big picture" in view to form coherent patterns out of what seemed chaotic to those who directly experienced small fragments of great events? Or is such post hoc patterning merely a product of the human impulse to impose order and ascribe meaning, and only vaguely related to the intricate phenomenology of warfare, which lies immune to adequate portrayal? Is war essentially a variant of Gödel's theorem, a phenomenon with such a "large number of variables or objects [that it] might be impossible to solve?"[11]

That wide discrepancy of views of war is visible in the realm of film, including television dramatization and documentaries, because those media have had far more effect on public views of warfare since the 1930s than print journalism or military history.[12] The filtering and sanitizing of images to conform to estimates of public acceptability has been normal practice. In 1962, for example, close-combat scenes in *The Victors* that its director, Carl Foreman, an infantry combat veteran, included to depict the gruesome mayhem of close combat were excised when audiences found them upsetting. Such discontinuity of imagery was noted a generation later by a Marine Corps veteran of extensive combat service in the Central Pacific theater in World War II. After watching the John Wayne film *Sands of Iwo Jima,* he observed that "there was no way you depict something that horrible. . . . If they'd done all that really happened, no one would want to see it."[13] That helps to explain further why so many veterans have found memories of their experiences elusive, or very difficult to describe in any detail.[14]

While most popular culture images lie far from reality by design or default, they have, along with "serious" fiction, drama, and history, not only been the main conduit for presenting views of warfare to general audiences, but also to many policymakers and elites lacking such experience.[15] Distortion has been compounded by the frequent use of censorship to mask the raw face of war, often out of fear that a clear view might weaken political support, recruiting, and fighting spirit. The effects of such suppression are more complex than that logic suggests. In both World Wars, recognizing the gap between sanitized images and reality fed combatants' cynicism after they saw the difference between official versions and reality.

Concern about technology conveying the raw stuff of battle to the rear echelons predated the invention of motion pictures. In the late nineteenth century, some German officers feared that the newly invented telephone might spread anxiety and alarm from units locked in battle to higher commands. Nor have the many subsequent advances in communications technology resolved the debate over where the optimum vantage point lies in controlling military operations, at the "cutting edge," or at higher levels where fragments of the "big picture" are assembled and perceived. With that tension frequently heightened by lower echelons' awareness of the uneven exposure to risk and violence at various levels of the hierarchy, and by unrealistic or silly expectations of commanders and staffs, many military professionals have seen communications technology as a source of chaos as well as an antidote to it.

In World War I, that tension, augmented by military failures and political stresses, led to dissension and mutiny in several armies and navies, even in the relatively stable British army, in which frontline troops dubbed staff officers "gabardine swine." In an especially caustic view of such stratification, Jean Norton Cru argued that French army staffs in World War I saw "at times what does not exist," because of lower units' modifying reports, out of fear of being disciplined, to conform with higher commands' expectations as well as adhering to "dogmatic principles which all the facts contradict."[16] A similar sense of anticipation diverging from actual events in combat flavors the writings of Field Marshal Erwin Rommel, who saw the outcome of battle as a simple product of two opponents' plans, and held that "only rarely in history, has a battle gone completely according to the plan of either side," and then only because of a victor's overwhelming strength or a loser's ineptitude.[17]

While the tumult of battle has sometimes been portrayed vividly in military history, it has been less visible in more formal descriptions of military forces' organizational processes. The mechanics of the command-staff process have very rarely been described, and then only cursorily, conveying little or no sense of detailed staff work. That is partly because of the fact that as headquarters personnel plot the "situation" on maps and in data bases, journals, and so on, they are unable to deal with all the minutiae flowing about them. As commanders and staffs grapple with overload and strive to sort the important from the trivial, their main concern is monitoring and maintaining the fighting capacity of the organization, not the amassing of records, although subordinate units sometimes sense otherwise. Here a special paradox arises because commanders and staffs need to have a deft sense of the condition of the subordinate elements; they are naturally concerned and empathic about conditions at the "cutting edge," but too great an awareness of that can impair their ability to function. As a result, they strive to balance detachment and concern while engaged in a blind struggle in which there is no clear format, or exact quantification of the risks and hazards which vary widely from situation to situation. In the early modern period, as noted previously, monarchs and generals were regularly under fire on the battlefield, but from the American Civil War to the present, top echelon ground, air, and naval commands have less and less shared the dangers faced by their contact elements, with the exception of some major naval battles and amphibious operations.

When that is considered in light of chaos-complexity-nonlinearity, it makes the sensing of seemingly minor details, including mood and morale, more important than previously thought. It raises the question of how much structure in itself at various degrees of separation injects critical degrees of filtering that occludes factors that might affect the complex equation of battle, far out of proportion to the magnitude suggested by quantity or linear comparison. That implicit tension between echelons has driven the debates over "mission-type orders" (*Auftragsbefehl*), and heightened the cleavage between leaders and led that arises from differences in privilege and safety. That presents a double bind in that leaders and staffs are expected to be sensitive to nuances and conditions in subelements, but to exercise command in a "hands-off" way, even if they believe they have information, perspective, or intuition that warrants their overriding the authority of subordinates.

While lower echelons' resentment toward what has seemed to be and sometimes was a lack of awareness or sympathy "up top," or "micromanagement," seems natural enough, it also reflects a misappreciation or denial of the fact that military-naval-air forces exist to do-or-die, literally. In western armed forces and their parent societies, however, countertrends have been under way since World War I, both in lowered casualty rates and in a dramatic increase in the survival of wounded. How much that has been a product of advances in medical practice *versus* the liberalization of political systems is not clear. The declining tolerance for heavy losses in war in the West was matched by increasing isolation of higher military commands from battle management, aside from the contingencies of nuclear war and terrorism. The disengagement of high echelons from fighting came into view during the Gulf War, when the question arose of deliberate targeting of military or political leaders.

Throughout the modern era, some commanders tried to be sparing of life, show concern for their forces' welfare, and eschew pomp and the display of power and personal privileges. Marlborough played the role of "Corporal John," although he was one of the bloodiest generals of his age. That pose, the "Nelson touch," and Omar Bradley's "G.I. general" image did not alter the fact that they, like military leaders of less populist mien, faced the contingency of having to send subordinates—sometimes friends, comrades, and relatives—into battle to face maiming, death, or captivity. That grim reality is still reflected in rigid disciplinary codes, hierarchical structures, and harsh military training and socialization in most military systems. Patterns were uneven, since reliance on coercion declined in both democracies and some authoritarian armed forces. From the mid-eighteenth century onward, the fading of the feudal aristocracy and the rise of populist values and nationalism, as well as of literacy and "thinking bayonets," generated such concepts as élan, bonding, camaraderie, cohesion, *Bluttkitt,* esprit de corps, charisma, and the "warrior spirit." All were based on the idea of developing inner motivation and enthusiasm in subordinates through positive rather than coercive leadership, and reliance on edicts and intimidation. In spite of such easing trends, which ran weakest in navies, all military leaders faced the irreducible reality of balancing affection and deadly purpose.

That dilemma was reflected in the rise of theories and schools of thought based on the application of psychology in leadership in several countries. Some of that trend was driven by the rise of "psychological warfare" and propaganda using mass media, and some of it arose from the search for techniques to identify skills and traits in selecting leaders and technicians, as well as airmen, submarine crews, signal personnel, and so on. Not only has that added an order of complexity to the waging of war, but it has also increased the potential for a sudden transition to chaos. Advances in command-and-control technology allowed subtle shifts in that delicate web of affection and control—in the mind of a commander, in a headquarters, or in subelements of a command net—to be quickly amplified and widely disseminated. The increase in the symbolic power of leaders was evident in World War II, both in the increase in personal security apparatus, and in the Allies' attempts to assassinate Rommel in 1941, unsuccessfully, and Reinhard Heydrich in 1942 and Yamamoto in 1943, successfully. Fear of a Nazi plot to kill key Allied commanders generated panic and paralysis during the Battle of the Bulge in 1944.

The emergence of such "maximum leaders" and the rise of modern dictators was mainly a product of technologies of communication, augmenting the centralizing of power, diffusion of images, and surveillance. Such networks, along with railways and then the internal combustion engine, allowed a massive expansion of the scale of military operations, and in the major wars of the industrial age, vast numbers of humans were fused with machines, linked by electrification, and extended over great regions of the globe. The view of warfare being intrinsically chaotic also grew along with that, paced by growing awareness that advantages gained by linking nodes and conveying messages quickly could be offset by such side effects. A question that complexity poses for military historians is to what extent the apparent chaos of war in a specific case, like the collapse of France in 1940, was a product of an array of specific prestates or catalysts that may not be fully tractable to analysis that permits a clear and certain conclusion.

Long before the "squad-leader-in-the-sky" syndrome in the Vietnam War, higher commanders overrode lower echelons. Electrical communication networks allowed higher authorities to issue edicts to commanders in the field, from the Crimean War through Bataan and Stalingrad to Korea, Vietnam, and the Gulf War. With that extension of war in space and the collapse of time in command-and-control, the removal of commanders from the zone of battle created the dilemma of commanders "over-identifying" with subordinates. That phenomenon, dramatized fictionally in Bierne Lay's *Twelve O'Clock High* and C.S. Forester's *The General,* has remained relatively marginal in military history, theory, and doctrine. Nor has it been as visible as it should be in recent analyses of cohesion and bonding of fighting men, since it lies at the heart of that crucial fusion. However intense that emotional enmeshing of leaders and led, and whatever its roots are, it does not fundamentally alter the fact that in war some are formally empowered to send others to risk danger or death, and to punish them if they refuse to do so. Such emotional crosscurrents and ambiguities cut across popular images of military authority being relatively neat pyramidal ordering. Whatever the subtler and more complex aspects of bonding and cohesion, and the relations between leader and led in war, that intrinsic organizational stress adds to the complexity and chaos of war, and to the difficulty, for historians at least, of determining just how that process was working in a specific case.

Contrary to that implicit intricacy and subtlety, much command literature focuses on organizational and technical dimensions, and on linear processes and systems aimed at imposing order and control. Divergent, nonrational aspects of leadership are sometimes dealt with by broad assumptions and allusions, or outright omission. The turbulence in the minds of leaders and others tends to be out of focus or invisible, because of the extreme difficulty of measuring and mapping, and the imperfections of monitoring and recording that chaos. Command process, if it is dealt with at all in military history, is usually portrayed somewhat vaguely as a backdrop of planning and orders, with few specifics about the actual flow of processional sequence, who is acting, or what is being done, beyond general references to plans and orders, and sometimes quotations. Command decisions made under pressure are seen as having some logical basis, sometimes out of an array of choices honed by the staff, and sometimes as the product of inspiration or intuition, but there is no Joycean stream of consciousness, beyond transcripts of voice transmissions. With that turbulent and intricate dynamic treated as a loosely defined "given," and commanders appearing as rational actors, error and failure in military operations are presented in most military history and biography as relatively clean-cut products of erroneous choice, flawed judgment, inadequate information, surprise, or faulty doctrine. Military and naval hierarchies proverbially seek to assign blame and responsibility in cases of failure. Little weight is given to the possibility that affairs were too complex to understand, at the time or even in hindsight, or that very small forces, choices, or turns of fate influenced great events.

The latter possibilities have been given greater weight in fictional portrayals of the ordeal of command such as *Twelve O'Clock High, The Bridge on the River Kwai, The General, Command Decision,* and *The Bridges at Toko-Ri.* Common in all of those is the commander's deep feeling for and protectiveness toward subordinates facing great danger. "Over-identification" is visible in each case, and tension between doctrine and operational reality is visible in the first four. So is the paradox that if higher commanders develop too much regard for those under their command, they may balk at or be unable to send them in harm's way, or may lose sight of their mission and even abandon it. In the first three, the fictional leaders, in trying to share dangers with those under their command, abandon the role of detached, impersonal decisionmaker at the apex of a command network. In the other two, the leaders agonize over at being unable to do so as they follow their duty, conforming to a doctrine, which, for all of its flaws, is seen as holding forth the promise of victory.

There has been relatively little analysis of military doctrine or style being antithetical to stratagem. Yet that is visible in the basic images and postures of soldiery, portrayed as hard, fierce, to and poised for battle in popular cultural imagery, military history, folklore, and literature, until photography showed how far reality often was from those models. Nevertheless, throughout military history, commanders, at all levels, from the Roman *miles gloriosus* to Shakespeare's soldiery, and from Prussian field marshals to P.C. Wren's Foreign Legion sergeants, struck stern poses. Even commanders of gentle or shy disposition, like Grant and Patton, cultivated a reputation for being willing and even eager to use brute force. Since war is about fighting and killing, that is hardly surprising, but those commanders presented another face of war, that of craftiness, dramatized by the contrast between the brave, fierce Achilles and the wily Odysseus in the *Iliad*. Reflection, stealth, and cleverness have generally been viewed as weaknesses in western military subcultures, and linked to effeteness, effeminacy, or cowardice. In World War II, sharp contrasts were drawn between the styles of Patton and Field Marshal Bernard L. Montgomery in the European Theater, and between Admirals William "Bull" Halsey and Raymond Spruance in the Pacific. At the time, each of those dyads exemplified the polarity of, respectively, headlong boldness and cautious passivity. The latter was viewed negatively, although perspective on both pairs had changed among military historians and professionals a generation later.[18]

Again, perceptions by lower-echelon fighting units that superiors failed to take their plight into account was sometimes correct,[19] and based on overlooking the fact that such empathy might impair a commander's ability to act, and might increase the risk and misery of subordinates. The roots of commanders' apparent heedlessness or callousness in military history are no easier to trace than other tendrils of causation. Among the more obvious negative influences are fear, fatigue, stress, and the proverbial "headquarters syndrome." Some leaders' distancing has been a deliberate attempt to avoid painful distraction. Neither Grant nor Haig, for example, who were each the first of their respective armies to command truly vast forces continuously engaged along a broad front, visited field hospitals, and Hitler ordered his railway car's blinds drawn when hospital trains came alongside. While it easy enough to ascribe those acts to callousness, or even pathology in Hitler's case, they may also have been closer to being the normal reflexes of any leader whose forces were caught in the grip of prolonged, unrelenting slaughter.[20] Some perspective on those cases is offered by Patton's loss of self-control when visiting field hospitals, and the famous incident in which he struck a soldier afflicted by combat fatigue.

Where does that inner, private, nearly inaccessible world of leaders' anguish and stress mesh with doctrine or history? One can speculate about seemingly eccentric behavior and psychiatric models of coping with tension by resorting to set forms and routines. In that context, aside from being a design for dealing with specific anticipated tasks, doctrine is like religious dogma in offering a kind of comforting ritual to lean on. That offers some explanation of why commanders and political leaders have so often fixated on it as a kind of psychic prosthetic to take them through dark valleys of doubt and anguish and toward uncertain futures, especially when under pressures that allow them little or no chance to change their reflexes. Here again there is no clear correlation between adhering to a course or changing it, and the ultimate outcome. In keeping with the Prussians' "Law of the Situation," each has led to success in some cases and to failure in others. Sorting out tangled connections between doctrine, operations, and individual actors' states of mind may become more tractable with the passing of time and the development of more sophisticated methodologies, but it has been out of reach so far, however confident the tone of those purporting to have seen it all, in battle or in crafting history.

Chaos in war has frequently been dealt with as a general attribute or characteristic of battle, and rather less evident atop command pyramids, in spite of many cases of leaders acting impulsively or strangely, such as going forward to seek personal involvement in combat, or fleeing. Since either of those practices can leave a command system without a designated decisionmaker, effects may be nearly indistinguishable. That obviously depends on how much a commander commands as opposed to being a kind of totem of authority without true function. Military memoirs and biographies offer little sense of that quality and only some fragmentary views of stress reaction, and dimensions of chaos and complexity at the "micro" level of military dynamics, let alone the actual workings of the minds of commanders. That seems likely to remain in the realm of imagination, artful and otherwise. Quirkiness and unpredictability of leaders, such as the eccentricities of Stonewall Jackson and Patton, have often been seen as a positive traits, especially when things seemed to work out. Anecdotes and impressions aside, the faint tracings of pathology are hard to track and evaluate, which, parochialism and defensiveness aside, helps explain why that dimension is more often portrayed in fiction than in history, or other military analytics.

Organizational analysts and militarily oriented social scientists have not delved too far into the anatomy of command behavior, aside from such doughty sallies as those of Norman Dixon in *The Psychology of Military Incompetence,* and Generals Bidwell and Richardson.[21] Few scholars examining military matters over the last generation have been as tough on the "brass" as journalists, or even former professionals. As noted earlier, military historians and defense analysts have tended to see rational deliberation and action and methodical staff work as givens, whereas informal group process and inner tensions, anxieties, and motives lie out of view, not necessarily because evidence has been lacking, but because of what is selected, emphasized, or ignored. Although visualizing of war as a kind of board or computer game played by martial chess masters has tended to crowd out the complexities, some journalists and historians have sometimes noted how each headquarters and staff is a unique community, with human interactions in that dynamic further from view than in fiction. It is also widely recognized that personal friction and accommodation are pervasive in the human interactions in military forces, and aboard ships and aircraft, including factionalism and personality clashes.

As variations in structure and technology add further orders of complexity to the intricate variable equation of warfare, the ultimate effect is that doctrine, as it passes through many such filters and lenses in the circuitry of various military headquarters and suborganizations on the way to becoming praxis, emerges as a very different thing, as highly fractalized, as, for example, the variants of close air support that had "evolved" in each major theater of war by the end of World War II. How much were such divergences the products of group process, individual whim, or style in those headquarters as opposed to being caused by differences in terrain, climate, enemy practices, or the type and flow of supplies? Sorting out such tangles of causation adds to the frustrations of military analysts trying to build algorithms of military operations. Roughly analogous to the paradox of the unfolding uncertainties encountered by physicists in trying to map the precise inner workings of atoms,[22] those intricacies make doctrine, even at its best, seem like a musical score from which all organizational elements try to play their parts under the direction of the commander as conductor who stands high in the balcony, in very dim or flashing light, faced with a loud and rowdy audience in a concert hall rocked by an earthquake. Small wonder that adherence to the score has ranged so widely from case to case, from equivalents of Bach string quartets to Dizzy Gillespie's jam sessions.

Another source of dissonance and complexity in military operations and organizations is the differences of the experiences within armed forces. From ancient Rome to modern imperial wars, and in the World and Cold Wars, they were sometimes made up of a virtual crazy quilt of veterans of different wars and campaigns. Lying past the consequent misalignment of a sense of what war is are other cross-strains, such as regional, religious, and class differences, and career tracks. Thus the top command and staff positions in several modern armed services have been held by veterans of different wars and operations,[23] overt and covert, or by nonveterans, or those of visibly marginal military capacity who are promoted for political or personal reasons. As such variations act as lenses both in the shaping of doctrine and in its perception and execution, doctrine may be as much or more a product of its shapers' experience as a novel construct based on insight, inspiration, or careful analysis of history, or a reaction to a new line of technology or to a foe's tactics or weapons. Whatever the case, human attitudes, habits, and experiences are the templates through which it flows.

Does the intractability of such dimensions mean inevitably that both the opacity of the past and the unforeseeability of future events make doctrine marginal, irrelevant, or not much more than a calculated gamble or a fragile catechism? How much can shapers of doctrine compensate for the irreducible coefficents of misalignment and error? Obviously, battle experience gives military professionals some sense of confidence within a logic that certain skills and traits apply in other settings. Often, however, military history is full of cases of wrong lessons being learned and experiences proving to be irrelevant as false confidence blunted sensitivity to crucial differences, or obscured the need or the chance to adapt or innovate. The fates of Braddock and Custer are well known, but there are also cases of effective carryover, like the amphibious warfare staff officers and commanders with expertise gained in World War II who helped speed planning of the Inchon assault landing in Korea in 1950. On the other hand, in World War II, which many have seen as the best-run American war, only MacArthur among the top tier of U.S commanders had substantial previous combat experience.

Considering the effects of battle experience on commanders and staffs makes it more apparent that an experiential mixture adds to both complexity and chaos, and makes it easier to appreciate why it has been so very hard to determine how much warfare is generic and how much it is unique. Each war and campaign has a different "aura," including differences in working vocabulary, technology, training, operational setting, and personalities, as well as politics, culture, climate and terrain of the region, and individual stress, fatigue, and fear. Those and many other elements not precisely measured, measurable, or perceivable at all make up a murky concoction. Once again, to what extent do such mélanges constitute complexities that are at least theoretically tractable as opposed to being chaotic and "beyond human ken"?

Further differences in war experiences stem from variations in types of wars themselves. What are the effects when doctrine, policy, and operations are controlled by veterans of large or small wars, of both, none, or of a mix of conflicts? How significant is the recentness of war experience, and the percentage of an armed force that it involves? Does success really tend to breed success, and failure breed failure, or is there an inverse pattern, or none at all? Although rarely considered as an aspect of complexity, let alone chaos, such experiential variety presents a dilemma with more than two horns. Even simple mixes have been complex in their effects. Like American colonial forces in the Revolution, the Israeli armed forces in the first generation of nationhood suffered from tensions between those with different military backgrounds in their officer corps.

That single dimension of complexity teased out of a very intricate nexus demonstrates the kinds of complex forces that affect doctrine as it "plays through" military systems, and their intractability to precise measurement. Naval historian Arthur Marder caught the essence of that indeterminacy when he asked rhetorically: "How important in judging the results of battle are the materiel losses and the casualties?" and answered, "in themselves, not very," drawing on the analogy of chess, in which victory goes to whomsoever checkmates, not to the captor of the most pieces.[24] Marder's metaphor, in dramatizing discontinuities in warfare, raises the question of how much an identifying of details is necessary to explain effect meaningfully, providing they are attainable.

To belabor the obvious, while historians and analysts have the advantages of hindsight and freedom from responsibility and urgency in making their determinations, that is not the case in operations. Commanders and staffs, and especially those in the intelligence "business," grapple with ambiguity and uncertainty amid stress and turbulence, destruction and horror. It is not surprising, then, since appraisals of what was about to happen or what had happened at the time were later found to lie far from those estimates, that such chronic discrepancy has contributed to a general sense of warfare-as-chaos. In the midst of battle, many details are invisible or blurred that later come into focus, but some do not, even after the passing of time. As a result, participants and historians are often unsure of what really happened, or why, and sometimes where or when, in spite of the advances in communications and information-processing technology, which have sometimes added to the sense of confusion rather than helping commanders and staff gain a clearer sense of what is going on. This confusion may be a product of direct destruction, jamming, or spoofing of communications systems in battle, of organizational compartmentation, or of isolation. It is not a new problem. Difficulties offsetting advantages arising from the adaptation of telegraph and railway networks were noted by military professionals in the mid-nineteenth century. In 1857, the German general staff formed a special section to deal with such problems, and half a century later, a vastly more complex universe had arisen from the meshing of transoceanic cables, telephones, and the first radios. The fusion of those inventions with mechanized and aerial warfare in World War I created further complexity.

In spite of that ongoing technical advance, combatants, in battle or afterward, have often been unable to see how their actions meshed with parallel or derivative events, or were related to the outcome. Of course, there is rarely any functional reason for them to have that kind of information, aside from the morale-boosting effects of having the big picture in view. Such knowledge may produce some hazard if it comes into the view of an adversary's intelligence system, although some have argued that the overall concept of higher-echelon commanders should be known to those whose task it is to carry out a fragment of such designs, since the diversion of events from expectations and disjunctures in communications between echelons in combat has so frequently confounded plans and expectations.

Just as those waves of technological evolution did not yield proportional coherence, the tone of surety of military professionals and historians has not been aligned with flawed perception and comprehension. Uncertainty about the sequence of events, unit and individual actions, the flow of orders, oversights, bad luck, and so on have fueled debates over many battles and engagements, major and minor. Since World War II, communications between echelons have been near constant and near instantaneous, as in air defense networks and air support and artillery fire control, and the radius of control has ranged from a few yards to many thousands of miles. (There is still a substantial lag in certain types of special operations.) In spite of all those developments, commanders' and staffs' ability to perceive battle dynamics has not kept pace with advances in communications technology, and some have sensed it to be an inverse relationship. This is in no small part because of the many parallel tracks of technical evolution and resultant increases in scale, velocity, and complexity that offset the apparent advantages of increasingly sophisticated command-and-control technologies considered out of context.

Recognizing the problems of overload and congestion in large communications nets in battle, on the eve of World War II, the British Army formed a special unit, the G.H.Q. Liaison Regiment, code-named "Phantom," whose detachments roamed forward in battle zones in vehicles and transmitted battle reports directly to high-level headquarters.[25] Phantom was a way of coping with the ambiguities and complexities that Sir Stafford Beer saw as common to industry and war, including defective intelligence, confounded plans, equipment failures, and differences in human behavior, "root causes" of chronic "failure to predict the manifold interactions of many variables in practice. . . ."[26] Since World War II, battle-monitoring techniques have converged toward such highly focused control systems as chemical processing, air traffic control, materials testing, accident investigations, and operational analyses, although defense analytics and military history have fallen behind, both in descriptive mechanics and in the tracing of dilemmas and paradoxes. Such analyses are very expensive and time-consuming under the best of conditions in hindsight, but in battle, there is little concern for, and less opportunity to trace, all that is happening, let alone to try to appraise motives and causes behind those activities.

While the increasing capacity to store and compare data seems to be bringing that problem under a greater degree of tractability, a blurred sense of what is happening has persisted at higher levels in operations and among historians trying to trace the flow of such events later. Concerted attempts have been made to carefully dissect operations, when incentives were strong enough and conditions allowed. Perhaps the most famous of such efforts was the Strategic Bombing Surveys conducted after World War II. In those and other operational postmortems, substantial effort and resources applied did not yield proportional clarity. Those involved in the U.S. Air Force's post-Vietnam CORONA HARVEST air power study found that, in spite of several major World War II studies, "no one had ever attempted to establish any measurable criteria for judging the successful accomplishment of an aerial mission." Lacking methodological sophistication, evaluators relied on their own and others' personal experiences and impressions, leading in turn to a wide fan of interpretations and arguments.[27]

When other operational postmortems fell short of clarity, those conducting them conveyed their sense of uncertainty regarding sequence and pattern with such ambiguities as "minor mischances and pitfalls of battle,"[28] or deeming impossible the construction of "a diagram [of Jutland] showing the whole of the battle."[29] A similar tone of vagueness, albeit unashamed, ran through both Rommel's accounts of his infantry actions in World War I, and of commanding an armored division in the dash across France in 1940. Both described his striving to impose some degree of order, sometimes very minimally, on the chaotic. Rommel noted that some operations closely followed plans and expectations and some did not, thus placing a premium on commanders' quick grasp of and reaction to sudden changes. As a junior officer, he frequently found himself in situations where "confusion reigned,"[30] or acting without knowledge of the situation.[31]

A generation later, as commanding general of 7th Panzer Division, Rommel encountered similar uncertainties and confusions in spite of his senior rank and access to better communications. In the race across France in May 1940, he met constant confusion, delays, units acting out of phase, "inadequate maps," communications failures, "wild confusion," and at one point confessed, "I had not any idea where the main body of the division was."[32] His accounts of battle resembled the "discontinuities" in information flow in the Battle of the Somme in 1916 described by John Keegan in *The Face of Battle* that "made the management of a battle, in . . . [a] tactile and instantaneous fashion . . . impossible. . . ."[33]

Rommel's descriptions of grappling with complexity and disorder conformed in spirit to Eisenhower's admission that he had made several major decisions without basis in certainty when doing "things that were so risky as to be almost crazy."[34] A similar sense of acting in confusion flavored a junior Soviet artillery officer's account of a desperate attack to break out of a pocket, after his loosely formed provisional force headquarters was "utterly ignorant as to the whereabouts of the enemy's forces and our own" for three days, in a success that amazed the Russians as much as the Germans they overran.[35] Such uncertainty and turmoil in battle has sometimes generated forms resembling the more abstract states of nonlinearity and chaos, such as the fractalization of British army tactics in the Boer War described by J.F.C. Fuller. Adapting to the free-roaming bands of mounted Boer guerrillas, as the British dispersed their large formations into small units, "Masses of men were split up into small columns, often so small as to be beneath the dignity of a Generals' command . . . and . . . consequently led by regimental officers—majors and mere captains, men not grown grey in peace, but become alive in war."[36]

A generation later, toward the end of World War I, fractal patterning was also evident in several tactical adaptations to the new technologies of war, including submarine warfare, aerial melees, and infiltration tactics. All of those diffuse configurations were attempts, at widely separate levels of elegance and forethought, to gain advantage over more contiguous, slower, and bloclike forces through dispersion and reduction of scale. That trend, visible in tactical thinking between the World Wars in the theories of J.F.C. Fuller and B.H. Liddell Hart in Britain, and of Felix Steiner in Germany, reappeared in World War II in submarine and antisubmarine warfare, in the blitzkrieg campaigns during the first third of the war in Europe, in aerial warfare, and in special operations. Such divergences in tactical thinking yielded substantial but not universal success, inasmuch as dispersal of effort, as an innovation, generated antitheses. In World War II, diffusion prevailed in some cases, but mass and concentration became dominant principles in the American amphibious warfare in the Pacific, Mediterranean, and Europe, in the Allied strategic bombing offensives, and in both Soviet and western Allied combined arms tactics, negating the Axis successes gained by freewheeling tactics in the first half of the war.

A senior Luftwaffe commander, General von Richtofen, saw that countercurrent at Stalingrad in late 1942, deeming the Wehrmacht's fluid mobile warfare a cresting wave, a phase-transition that was also visible in North Africa. Recognizing a doctrinal vacuum, he complained that there was a lack of "some clear thinking and a well-defined primary objective. It's quite useless to 'muck about' here, there and everywhere as we are doing."[37] In calling for a *Schwerpunkt*—the defining of a crucial point of decision—amid the mounting chaos on the Volga, von Richtofen saw a need to impose order on complexity, much in the spirit of the 4th-century B.C. Chinese general Sun Pin's aphorism: "One who wishes to unravel the confused and entangled does not grasp the entire skein. Strike at a salient or unprotected place."[38] A similar tone of confusion marked other accounts of the Stalingrad campaign. Colonel General Kurt Zeitzler, chief of the German army general staff at that time, writing fourteen years later, could not say how many German troops were trapped there. Resorting to lyricism and references to chaos, turbulence, and indescribability found in many martial memoirs and much military history, he invoked such lurid and vague metaphors as "a nightmare without end" as a description of how "whole formations melted away [and] the Sixth Army was consumed by a fire until all that was left was slag."[39] Such language appeared in other accounts, reflecting how, from 1942 to 1945, the Wehrmacht confronted not only growing Allied strength, but forces deployed against them with a much closer linking of doctrine to strategy, and under a more literal and pragmatic view. Other sources of the Germans' sense of chaos were the army's extremely intricate hierarchical organization,[40] and the Wehrmacht's relying on technological nostrums and structural reshuffling to compensate for doctrinal failure and lack of coherent strategy.

Command architecture and process as sources of friction and chaos in war can be glimpsed in other accounts, factual and fictional, but it is hard to gauge their effect precisely because of a dearth of details on personal motive and informal processes, and due to the limits of methodology, and of language and graphics in describing battles. Narrative explanation and highly abstracted maps remain the principal format of military history, backed up by statistics and illustrations. Both war planners and military historians have tended to see the flow of military operations, however complex, conforming to the intent, generally if not in detail, of leaders on the winning side. However logical those assumptions may seem, the wide variances in personal accounts and evidence suggest otherwise. The pervasive, brutal randomness of warfare, the widespread "collateral damage" and slaughter of innocents, muddling through, blunders, and accidents of war stand in tension with the image of war as a chesslike contest based on purposive order, and somewhat closer to the convergence of tidal waves. Nevertheless, in the dynamics of war portrayed in journalism and history, that virtual swirl of forces and events are seen as totaling up to a clear sum of victory-and-defeat.

Despite such overlooking of ambiguity, it has not been possible—despite Soviet claims that their concept of "correlation of forces" has a viable doctrinal algorithm—to determine the critical components of an equation of force in numbers, technical quality, tactical skill, genius, or less tangible elements such as morale and training. Theoretically, being able to do that would ensure the outcome of a specific operation in advance—unless both opponents relied on such a logic in configuring their forces, thus generating a further magnitude of complexity while confounding their search for controllable patterns. That helps explain Wellington's view that describing the flow of battle is no more possible for historians than it is for commanders. Standing away from such manifold details has often been seen as a way to gain sharper perspective, that is, excluding or reducing chaos and complexity to reach general essence, clarify perception, or help in designing refined concepts and methods. To the extent that the application of logic is relevant it confronts the aphorism that all generalizations are false, which leads on to the dead end of begging the question. Because nuances have been discarded or overlooked does not mean that such subtle, invisible, or immeasurable forces have not significantly influenced the flow of history.

Since absolute determination of that lies out of reach, sensing and measuring, let alone the controlling of even relatively low levels of complexity, remains more art than science. Some sense of that is suggested in the many variants that can be spun out of the simple model of types of wars devised by F. H. Hinsley, a historian of statecraft and intelligence processes:[41]

> One war may be almost entirely due to the given conditions and hardly at all the consequences of the conduct of the men involved. Another may be almost entirely due to that conduct and hardly at all to the consequences of the given condition.

Between the bounds of that simple polarity lie many combinations of those two elements. Folding in all the causal and shaping elements found in histories of warfare produces a conglomerate that far exceeds the criterion of complexity set by William Livingston, who, in a corollary to Ross Ashby's law of requisite variety, held that something is "complex when it exceeds the capacity of a single individual to understand it sufficiently to exercise effective control regardless of the resources placed at his control."[42] In the future, applying new techniques of analysis to the study of warfare may bring more of those complex elements into view, and help historians and military professionals comprehend how and why events transcend the understanding of participants at the time. Analysts would at least gain a clearer view of the boundary between attainable precision and irreducible ambiguities and uncertainties. It remains to be seen if—or how much—gaining such a perspective on war dynamics yields fuller understanding. In seeking enhanced resolution of combat and war, military professionals and historians both resemble engineers studying bridge failures who find their attempts to reconstruct events "too complex for complete computer analysis," since they "require a host of simplifying assumptions" that "seldom match behavior in the field."[43]

Considering the causes of chaos in warfare has another edge to it. Commanders since ancient times, including Hannibal, Genghis Khan, and Napoleon, have used various ploys to engender panic among their foes. The impact of rapid increases in technical evolution in the late nineteen and early twentieth centuries of war increased the view of it as frustratingly complex and chaotic, or as chaos itself. World War I saw a further bounding forward in tactics and methods aimed at inducing chaos, from the concentration of carnage in massive, prolonged artillery barrages to aerial bombardments, gas, and tank attacks, whose effect was proportionally more psychological than physical. In the final phase, tactics were designed to disorient and intimidate enemy forces with relatively little physical destruction (e.g., the Australian "peaceful penetration" trench raids, the Germans' infiltration tactics, and the British revving tank engines behind the front to panic German troops).

Major psychological shocks were generated by the sudden appearance of "secret weapons" during World War I, such as various types of poison gas, flamethrowers, tanks, and the Paris gun. Those, along with the epidemic of mutinies, increased sensitivity to the potential of using weapons to intimidate as well as damage or kill in exclusively physical terms. With that effect in view, just after World War I, a German military critic lamented that failing to use bomber aircraft against the French main supply route into Verdun in 1916 lost a chance to create a "chaos that could not be disentangled."[44] In much the same vein, in the mid-1930s, J.F.C. Fuller expressed the hopeful view of many advocates of armored and aerial warfare in predicting that "as the warfare of machines is less destructive of property, and more destructive to will and nerves, the tactical object of war will be demoralization and disorganization rather than destruction and annihilation."[45]

Although bloody attrition ultimately marked much of the fighting in the Far East and Russia, there were also some countertendencies. Casualties were relatively low in the major armored drives of 1939–41 and 1944, and a host of terrorist, guerrilla, "special," and psychological warfare tactics appeared, based on the use of small, mobile forces to avoid direct linear combat. Some were successful against heavier, more numerous units, keeping adversaries off balance and forcing them to allocate disproportionate resources to the contingency of fending them off. Noel Koch pointed out the logic of chaos underlying terrorism in observing, "There's something tidy about our wanting to put a name to everything that happens. But if you want to create anxiety and unease, a clever way of doing it is to let there be no reason for it."[46] Inducing chaos and panic was also a central element in the Soviet concept of "reflex control" and related paradigms. In the then-classified Soviet military journal *Voyennaya Mysl'*, an essayist, in discussing the psychological effects of surprise, listed "impulsive chaotic actions" as a primary result of being taken unawares.[47] In a lengthy exegesis on these subjects in Russian defense-related and scientific periodicals and books from the mid-1970s to the mid-1980s, the Soviet military appeared to be closer to enfolding chaos-induction and nonlinearity into military doctrine than any nation's armed forces had since the *Reichswehr,* and the Wehrmacht of early World War II.

Such deliberate generating of chaos in adversaries' systems was attempted by other nations at other times[48] and is the underlying logic of using and brandishing force across the spectrum of conflict, from propaganda and psychological warfare to nuclear strategic theory. Not only has chaos been seen as an implicit aspect of combat and war, but the Romans viewed war in itself as a type of chaos, and sometimes as its ultimate embodiment, "actually determining whether chaos would triumph in the universe."[49] As a loose allegory, that throws at least a dim light on why military doctrines and plans have so often been proved to be out of phase with reality.

NOTES

1. I.S. Konev, *Year of Victory* (Moscow: Progress Publishers, 1984), p. 240.
2. Albert Parry, *Russian Cavalcade* (New York: Ives Washburn, 1944), p. 75.
3. For several personal accounts in a single anthology that convey a sense of chaos, see Edward K. Eckert, ed., *In War and Peace: An American Military History Anthology* (Belmont, Calif.: Wadsworth Publishing Co., 1990), pp. 58, 128, 147–48, 163, and 357.
4. See Roger Spiller's essay on such effects: "The Price of Valor," *Military History Quarterly,* 5:3 (Spring 1993), pp. 106–110.
5. Lee G. Miller, *The Story of Ernie Pyle* (New York: Viking, 1950), p. 277.
6. John Ellis, *The Sharp End: The Fighting Man in World War II* (New York: Charles Scribner's Sons, 1980), pp. 107–112.
7. An example from the American Civil War is David W. Bight, ed., *When This Cruel War Is Over: The Civil War Letters of Charles Harvey Brewster* (Amherst: University of Massachusetts Press, 1992), p. 312.
8. Bodo Zimmermann, "France, 1944," in Seymour Friedin & William Richardson, eds., *The Fatal Decisions,* trans. Constantine Fitzgibbon (New York: William Sloane, 1956), p. 225.
9. Bruce H. Norton, *Force Recon Diary* (New York: Ivy Books, 1991), p. 252.
10. Robert A. Moore, quoted in Bob Tutt, "World War II Veterans Recall Bloody Battles at Kasserine Pass," *Houston Chronicle,* February 14, 1993, p. 33A.
11. Joseph Traub & Henryk Wozniakowski, "Breaking Intractability," *Scientific American,* 270:1 (January 1994), p. 107.
12. E.g., see Richard Holmes, *Acts of War: The Behavior of Men in Battle* (New York: Free Press, 1985), pp. 66–69.
13. Keith Wells, quoted by Bill Whitaker, "Aggie Vet Portrayed by John Wayne Dislikes Film Version of War," *Bryan-College Station Eagle,* September 24, 1991, p. 2A.
14. For a discussion of this phenomenon, see Roger Spiller, "Shell Shock," *American Heritage,* 41:3 (May–June 1990), pp. 75–87.

15. E.g., see Roger Beaumont, "Images of War: Films as Documentary History," *Military Affairs*, 35:1 (February 1971), pp. 5–7.

16. Jean Norton Cru, *War Books: A Study in Historical Criticism*, ed. & trans. Stanley J. Pincentl (San Diego: San Diego State University Press, 1988), p. 11.

17. B.H. Liddell Hart, ed., *The Rommel Papers* (New York: Harcourt, Brace & Co., 1953), p. 519.

18. For perspective on Montgomery's style as a function of Britain's lean resources, see Russell F. Weigley, *Eisenhower's Lieutenants: The Campaign of France and Germany, 1944–1945*, Vol. 1 (Bloomington: University of Indiana Press, 1981), pp. 74–75; for a recasting of the roles of Halsey and Spruance, see Ken Hagan, *This People's Navy: The Making of American Sea Power* (New York: Free Press, 1991), pp. 320–328.

19. E.g., the versions of nicknames based on "butcher" awarded to Air Marshal Sir Arthur Harris and General Charles Mangin.

20. See John Keegan's comments on S.L.A. Marshall in *The Face of Battle* (New York: Viking Press, 1976), pp. 70–72.

21. E.g., see F.M. Richardson, *Fighting Spirit: A Study of Psychological Factors in War* (London: Leo Cooper, 1978).

22. E.g., see Donald N. McCloskey, "History, Differential Equations, and the Problem of Narration," *History and Theory*, 21:1 (1991), pp. 1–20.

23. E.g., see Al Santoli, *Leading the Way: How Vietnam Veterans Rebuilt the U.S. Military: An Oral History* (New York: Ballantine Books, 1993).

24. Arthur J. Marder, *From the Dreadnought to Scapa Flow: The Royal Navy in the Fisher Era, 1904–1916* (London: Oxford University Press, 1966), p. 206.

25. An account of the unit's formation and activities is R.J.T. Hill, *Phantom Was There* (London: E. Arnold, 1951).

26. Sir Stafford Beer, *Decision and Control* (London: John Wiley & Sons, 1966), pp. 38–39.

27. Robert F. Futtrell, "Commentary," in William Geffen, ed., *Command and Commanders in Modern Military History* (Washington, D.C.: Office of Air Force History, 1971), pp. 284–285.

28. C.E. Lucas-Phillips, *Alamein* (Boston: Little Brown, 1962), p. 296.

29. W.S. Chalmers, *The Life and Letters of David, Earl Beatty* (London: Hodder & Stoughton, 1951), p. 357.

30. Erwin Rommel, *Attacks* (Vienna, Va.: Athena Press, 1979), p. 193.

31. *Ibid.*, p. 30.

32 Liddell Hart, ed., *The Rommel Papers*, pp. 55–57.

33. John Keegan, *The Face of Battle* (New York: Viking Press, 1976), p. 262.

34. Alfred D. Chandler, ed., *The Papers of Dwight David Eisenhower: The War Years*, Vol. 3 (Baltimore: Johns Hopkins University Press, 1970), p. 1712.

35. Fred Virski, *My Life in the Red Army* (New York: Macmillan, 1949), pp. 159–160.

36. J.F.C. Fuller, *The Army in My Time* (London: Rich & Cowan, 1935), p. 88.

37. General Freiherr von Richtofen, in Walter Goerlitz, *Paulus and Stalingrad* (New York: Citadel Press, 1963), p. 192.

38. Quoted in Sun Tzu, *The Art of War*, trans. Samuel B. Griffith (London: Oxford, 1963), p. 60.

39. Kurt Zeitzler, "Stalingrad," in Freiden & Richardson, *Fatal Decisions*, pp. 169 and 183.

40. Alexander Kluge, *The Battle,* trans. Leila Vennewitz (New York: McGraw-Hill, 1967), p. 110.

41. F.H. Hinsley, *Power and the Pursuit of Peace: Theory and Practice in the History of Relations Between States* (Cambridge: Cambridge University Press, 1963), p. 331.

42. William L. Livingston, *The New Plague: Organizations in Complexity* (Bayside, N.Y.: F.E.S. Limited Publishers, 1985), pp. 1–11.

43. Kenneth F. Dunker and Basile G. Rabbat, "How Bridges Fail," *Scientific American,* 266:3 (March 1991), p. 68.

44. Horne, *The Price of Glory,* p. 208.

45. Fuller, *Army in My Time,* p. 172.

46. Quoted in Douglas Jehl, "Americans Feel Terror's Senseless Logic," *New York Times,* March 1, 1993, Sec. 4, p. 1.

47. Z. Paleski, "Psychological Aspects of Surprise," *Voyennaya Mysl',* No. 7 (July 1971), p. 93.

48. E.g., see Douglas Robinson, *The Zeppelin in Combat* (London: G.T. Foulis, 1962), pp. 25 ff.; and in the British Public Record Office, AIR 1/459, 1/520, 790, and 2295.

49. Sue Mansfield, *The Gestalts of War* (New York: Dial Press, 1982), p. 69.

6

The Ordering Impulse

For the last two centuries, technology has enhanced the ability of military leaders to extend their control over military operations while adding to both the destructiveness, and the sense of randomness and disorder in warfare. That is hardly surprising, since in warfare, the impulse to mayhem and consequent chaos is given its greatest outlet, in a peculiar fusion with also-strong human drives toward order. While most damage wrought by war has been beyond battlefields, and that wanton destruction, atrocity, pillage, and rape stand in sharp contrast with the ritualized linearity of military drill, discipline, military codes of honor, and tactical formats, they are not portrayed in actual proportion to their frequency. Beyond the fact that many such activities are usually carried out furtively, even when there is substantive data about side effects and "collateral damage," they are usually not evident on battle maps, operational or historical, and are not included in the description of combat and supporting statistics.

Overall, military historical narrative and accompanying apparatus tend to soften the sense of tantrum and mayhem. In keeping with that disposition to discern order amid chaos and complexity, military historians have tended to take war and military institutions on their own terms. Few have viewed war as pathology or aberrance, even when recognizing the terrible destruction, although some who have sought to trace the roots of human combativeness—social scientists, bellometricians, pacifists, systems analysts, operational researchers, and polemologists—have nibbled around the edges of the terra incognita of warfare, like fourteenth-century cartographers. Wherever those efforts ultimately lead, the tentative and fragmentary attempts to frame an anatomy and etiology of human collective violence have so far only contributed to the widespread impression that warfare is implicitly chaotic.

Within military institutions, the cross-pull between order and chaos is visible in many forms, like the stereotyped social demeanor of the British, French, German, and Japanese officer corps, each a variation on the theme of imposing a behavioral format on disorder. Those stylized behaviors, like stays on a sailing ship's masts, have served as braces to support leaders as they struggle to maintain composure and control their own and others' behaviors under the stress of battle. Biography, journalistic accounts, and history occasionally present a glimpse of one or a few points on that shimmering frieze of myriad variation at the individual level, but such behavioral minutiae lie beyond historical description—and the discriminating capacity of command-and-control systems. Yet there is no serious disagreement among military professionals or historians about such micro factors being a source of complexity, crucial on occasion, or that the tapestry of warfare tends to the chaotic. Nagging questions that emerging here include the following: Are such seemingly minute and anomalous aspects of war tractable to rational analysis? Can ways be designed to monitor and control them?

The successes of British and American scientists in military problem-solving in the late 1930s and World War II—then known as "operational research"—led many managers and administrators, when commercial computers appeared in large numbers in the 1960s, to anticipate a virtual revolution in the ability of leaders to oversee and orchestrate complex business, bureaucratic, and military processes, and to devise optimum patterns of control. Those hopes were widely confounded during the Vietnam War[1]; by the late 1970s, several generations of computer evolution had allowed much closer focusing on systemic intricacies, and greater appreciation of how complexity and chaos had thwarted earlier optimistic expectations regarding a "managerial revolution." Those advances also made it possible to approach theoretical problems in non-linearity long recognized but left unaddressed due to limited calculating capacity.

The most obvious applications of nonlinear theoretics in the military realm were in the areas of command-and-control and communications, including cryptanalysis, as well as aerodynamics, tactics, and certain types of weapons effects. Growing interest in "fuzzy" and "squishy" aspects of systems dynamics not only bore out some commonsense perceptions and intuition, but helped explain why combatants,' journalists', scholars', and analysts' depictions of warfare fell well short of constituting a standard format of combat processes. Such viewing of warfare as increasingly chaotic is not new. From the Renaissance onward, admission of bourgeois into the officer corps of European armies was paralleled by a virtual rainbow of destabilizing elements, including a fractalizing of tactical forms, dispersion of military operations, and growing infusions of technology into the waging of war. Mourning of the passing of the old order was heard again and again, in Prussia after Jena, in America after the Civil War, and across Europe after World War I, as expressed so lyrically in Jean Renoir's film *Grand Illusion*. The deeper irony of the graciously archaic being displaced by greater crudity and destructiveness was reflected in J.F.C. Fuller's enthusiasm for mechanizing military systems and bringing proletarian mechanics and bourgeois engineers into the officer's mess, while also lamenting that the Boer War was "the last gentleman's war." In a similar vein, former cavalry officer Winston Churchill leveled his well-known blast against the "lights of perverted science" that he saw wiping away the last traces of martial glory, but most of the victims in the bombing of Germany that he presided over in World War II were civilians.

Since that happened in all strategic air campaigns, as boundaries between categories of targets became blurred, it is not easy to tell how much such "collateral damage" was a side-effect of a certain level of technical evolution and resultant increasing complexity, or a product of a fundamental human cruelty, masked in the sterile, bureaucratic guise of "policy" and by secrecy, but amplified by technology. Whatever the causes, the imprecise effects of weapons of mass destruction added degrees of chaos to the "business of war." Although traces of the old patterns of orders and chivalry remained visible in World War II, most notably in North Africa and in the air war in Europe, the overall trend was toward the abandoning of laws and customs of war observed widely, but not universally, since the late seventeenth century, when the Thirty Years' War's atrocities led to the founding of modern international law.

While such rules were frequently ignored in insurgency, guerrilla warfare, civil and colonial wars, and terrorism, relative constraint prevailed in Europe before World War I. As that struggle quickly became a "total war," like the Napoleonic and American Civil wars, it vastly diverged from most of the initiators' plans and expectations. Even before censorship was eased afterward, millions of noncombatants had been shocked by the scale of its horrors as mass-produced destructive technologies got out of control, creating vast, bloody chaos beyond the intent or control of military and political leaders. While many military professionals were also horrified, others glimpsed a road to tactical advantage by accepting some degree of disorder and having small units grope for weak spots in the enemy forces amid the vortices of battle, hence the German army's "infiltration" system and the British army's tank tactics of 1917–1918. Both, by yielding to the disorder of the matrix of war, offered a way to regain some degree of momentum and effect.

As further revulsion followed postwar revelations of stupidity and butchery, pacifism became a powerful force in the western democracies' political arenas. In spite of massive curtailing of armaments, by the mid-1930s, technical advances, most especially in aviation, had extended the threat of mass destruction from relatively confined battle zones to whole nations and regions. A plethora of science-fiction works projected the grisly realities of World War I and some of those apocalyptic scenarios of the futility of war, like the film version of H.G. Wells's *Things to Come* (1935), proved remarkably prescient. An end to warfare had been predicted at various points along the path of technical evolution (e.g., with the invention of dynamite and the machine gun). At each step, however, it was found that such advances did not make waging war impossible, but far more expensive and terrible, and no less likely.

In the realm of military praxis, the converging of technologies, such as the fusion of aircraft, radios, and tanks, yielded both advantages and dilemmas. As top brass became more remote from combat, they sought to deploy larger forces and wield authority although they were less and less able to assess conditions personally, or to exercise personal leadership. Terms like coup d'oeil (sweep of the eye) or *Fingerspitzengefühl* (fingertip sensitivity) fell more and more into the province of lower-echelon commanders. The isolation of high-echelon commanders from the physical dangers of war did not make them immune from its stresses as they strove to reimpose some degree of order on what appeared to have gone well out of control. The toll of the stress of coping with complexity can be seen in photographs of commanders, in the gravures of care etched into the faces of political and military leaders. Beyond those were commanders' nervous breakdowns, short tempers, and quirkiness, proof that Furmanov's dictum that there is no emotional calm in combat[2] extended outside the battle zone.

As gaps between operations and the points of control widened, some forms of combat were also changing. Submarine warfare initially caused a crazing of patterns in World War I as the dialogue between stealth and surveillance was drawn out to a fine edge in the war at sea at hundreds of separate points. Their advantage was initially blunted as the British countered with randomness in the form of Q-ships, and then reintroduced order into the tactical equation with convoys. Submarines usually tried to avoid direct combat with an armed enemy, instead attacking the unwary or defenseless, a pattern that continued in World War II, but which was modified in stages by aircraft, radar, sonar and a host of other refinements in technique. In 1940, the U-boats gained major successes in momentary collaboration with Luftwaffe long-range aircraft, and Admiral Karl Doenitz and his staff, using radios from French bases, orchestrated attacks on convoys by groups of submarines with varying degrees of success.

On land, in World War I, the major fronts became the grinding shed of gigantic industrial systems, and, like the submariners, many soldiers became less and less warriors and more engaged in exhausting, repetitious labor such as digging, hauling, and servicing thousands of guns, vehicles, and horses. When battle was joined, combat resembled a blind clash of mobs in a dense fog. Aside from snipers and some hand-to-hand trench fighting, most killing and destruction was depersonalized, and the main media of military power became mass artillery barrages and machine guns firing mechanically controlled or set patterns. The weapons of the infantry—trench mortars, grenades, and submachine guns designed as "trench cleaners"—were based on a real destruction, not the individual heroism depicted in fiction and in the heavily censored, fancifully illustrated press. The connotation of being "in combat" evolved from one of fighting hand-to-hand to having been at risk from weapons fired without direct aim, or mines or booby traps, or shells or bombs arriving from great distances, often without directly seeing an enemy. Dispersal, camouflage, diversification, and bureaucracy in the mass armies all added to the sense of complexity, disorder, and chaos, and the feeling of anonymity and randomness. Storm troops, like their successors in World War II, airborne, special and elite forces, and submariners, often avoided direct combat and relied on stealth and secrecy.

Such stresses and forces induced in those affected a chaotic sense separate from the actual state of affairs. In the spirit of Walter Bedell Smith's aphorism, "No one thinks clearly when he's scared," stress and fatigue impaired the coping capacity of those directing operations,[3] and also affected anticipation, prediction, and planning. In spite of that, and other complicating forces like vendor influence, political pressures, and the cryptic, impenetrable motives of individuals and factions, analysts have expressed surprise that leaders aware of rationalistic decision-making paradigms have nevertheless tended to rely on intuition, impulse, or nonrational logics.[4] Those proponents of logical constructs had no sense of the stresses affecting leaders as they tried to cope with a bewildering flood of data, and the not wholly rationalized concatenation of energies loosed in war, each wave of which generated, in turn, greater orders of complexity.[5]

The tangled system-states of war have lain far from constituting a clear, logical model of conflict resolution, tractable to rational control, even though many of the formats and ideals of such an ordering have endured, with less and less of the substance. By making war leaders' impulses and hunches seem apt and rational, military tradition, ritual, and history have maintained a vestige of the traditional role of kings as anointed warrior chieftains that died out only at the beginning of the twentieth century. Through World War I, monarchs served as commanders in the field, symbolically if not substantively. The royal function as the capstone of the legitimate social pyramid based on the role of war chief was the basis of the Prussian general staff model that emerged in the early nineteenth century, which was designed to serve kings heading their armies in war who were less martially gifted than Frederick the Great. The Prussian military professionals served under the fiction of being martial servants, performing administrative tasks seen as menial and maintaining forms of deference while gaining functional control by replacing royal intuition with their own command-staff and bureaucratic processes. Ironically, under Hitler, the general staff was pushed back toward being lackeys and arch-clerks, working out details and executing the intuitively based schemes of the master seer who, like ancient kings, reestablished the role of the national leader as visionary.

That role has had substantial momentum. The vast majority of modern dictators have posed as infallible, intuitive martial chiefs and worn military uniforms. Some, like Pilsudski, Ataturk, Petain, and Chiang Kai-shek, had solid credentials as field commanders, but most held no high military rank prior to gaining power through other means, including Stalin, Castro, Perón, the Shah of Iran, Qaddafi, Idi Amin, and Saddam Hussein. Such forms and behaviors are evidence of a deep-seated hunger to impose order, but they have also been mechanisms for generating disorder. Dictators seeking to prove their prowess as commanders sometimes generated literal chaos, an index of how much the role of commander has been seen as a matter of appearances rather than substantive competency. "Bearing" is listed as first among leadership traits at the United States Military Academy, making it a major dimension of the mystique arising from the command of armed forces,[6] and its link to social and political power. There is, of course, a wide range of styles, from the blustering of Custer, Allenby, and Patton to the calm comportment of Grant, Lee, Bradley, and Harold Alexander, or the folksiness of Zachary Taylor, and George Meade.

Beneath those veneers, commanders have struggled with both powerful organizational and inner psychological stresses. John Keegan has argued that generals must deeply fear the loss of control over their forces, and most especially the outbreak of panic,[7] mirrored in the mass flights of civilians in war and crisis.[8] Concern over such psychic disintegration of military forces in a rout, surrender, or paralysis has led some analysts to search for the "breakpoint," the crucial instant when a force in combat collapses or flees.[9] Yet as much as order is sought for its own sake in military socialization, training, organization, and practice, the onset of chaos in battle, from medieval melees to the airborne and amphibious operations of World War II, has often led to functional loss of initiative or loss of momentum.[10]

No firm aggregate of evidence points to an index of control or coherence clearly linked to winning or losing. It is not clear, for example, that conditions aboard John Paul Jones's victorious *Bon Homme Richard* were really less chaotic than those on the vanquished *Serapis*. Mechanization, aviation, and the dispersal of mass armies and air forces in the face of ever-greater firepower have rendered that aspect of warfare even less certain. The grappling of modern armies, fleets, and air forces has often resembled the mingling or collision of energized fluids, or the blend of gas clouds, phenomena vaguely sensed and describable only in general mathematical terms. That has made graphic depiction of battle dynamics increasingly difficult even as evolutions of computer technology allowed more rapid and exact portrayals of operations. Although that increase in resolving power appears to offer advantages in both command-and-control and historical analysis, it might also only yield another degree of abstraction, saturate the senses of practitioners and analysts, or prove useful only to one or the other.

Such visions of variegated minutiae making up the dynamics of battle are in keeping with proverbs about God—or the Devil—being in the details. The British scientist P.M.S. Blackett, considering the difficulty of tracking battle dynamics, saw that "success . . . in war in general and almost all air operations is due to the sum of a number of small victories, for each of which the chance of success in a small operation is small."[11] Beyond many complaints about the lack of useful models that encompass and adequately describe such details in the aggregate[12] lies a broader range of uncertainty about such complexities, including the Russian tradition of relying on broad numerical abstractions,[13] the various models drawn up in the spirit of Jomini's ponderings,[14] and the view of an operations researcher that "state variables are the numbers" in the analysis of combat.[15]

The mounting interest in nonlinearity among scientists from the mid-1970s onward and in intractable imprecisions in certain complex phenomena[16] was paralleled by constant increases in the magnitude of intricacy in various technologies, from computer chips to parallel processing. The improvements in display technologies were matched by an increase in the difficulty of visualizing the growth of what R.L. Gregory called "the world we cannot see,"[17] a problem in view well before the development of computer networks. In World War II, for example, a "pipeline" effect appeared as the U.S. armed forces expanded from less than 2 million to over 13 million in three years. Staff officers were stunned to find a vast, diffuse horde of people floating about in the system, not within units and commands, but traveling, sick, training, on temporary duty, on furlough, AWOL, replacements, and so on,[18] a proof of Hans von Seeckt's warning against seeing armies as fixed entities.[19]

From the early nineteenth century onward, the search for orderly patterns of warfare amid such swirling changes led to the framing of ever-longer lists of principles of war,[20] and such schema as Trevor DuPuy's "timeless verities of combat,"[21] B.H. Liddell Hart's "strategy of the indirect approach,"[22] John Wood's precepts of armored warfare,[23] and the tenets of the maneuver warfare enthusiasts of the so-called Reform Movement in U.S. defense circles following the Vietnam War. Such hungering for order persisted in the face of the skepticism of many military professionals who, like Rommel, saw military theoretics as "academic nonsense."[24] The search for guideposts encompassed the lofty strategic theorizing of von Schlieffen and, at the lowest levels of practice, what Liddell Hart deemed "pillars of fire by night," not only battle drills, but checklists, SOPs (standing operating procedures), formatted "estimates of the situation," and order and message forms. At a somewhat less mechanistic level were guidelines and rules of thumb that became clichés in military and naval parlance, like "fire-and-maneuver," "two-up-and-one-back," and "crossing the 'T'," as well as J.F.C. Fuller's "expanding torrent" model of mechanized breakthrough, Heinz Guderian's *Angriff Ohne Befehl* (attack if lacking orders), and various armies' trench warfare tactics in World War I.

Ironically, although attempts to rationalize technique were usually aimed at imposing order, some generated conceptual complexity in the form of debate and friction. Veteran commanders who became military historians and analysts often drew widely different conclusions from their experiences and from the study of war. For example, J.F.C. Fuller viewed differences in weapons technology as dominant over all other factors in yielding tactical advantage,[25] while David Hackworth, in the spirit of the samurai, argued that warriors' spirit transcended all.[26] In nonlinearity terms, such dichotomies constitute actual double-attractors, of which there have been many in military history and analysis. For example, in a variation of the "light-heavy" debate noted earlier naval officers, strategists, and designers since the eighteenth century have argued the merits of a fleet built around a core of heavy, powerful capital ships as opposed to one composed of fast, independent raiders. Paradoxically, those engaged in that *guerre de course* versus *guerre d'escadre* debate, in their striving to impose clarity and precision on the complexities of warfare, ended up generating further orders of uncertainty.

Little sense of such intricacies is conveyed in popular or academic military history, or of the elaborate division and fusion of labor, or formal and informal processes. That may be partly due to sparse and uneven evidence, or the fact that many would find it as tedious to read as historians would the crafting of it. In any case, the mystique of the decisive and dominant leader pervades military historiography, as does the sense of war being a contest of generally rationalistic competitors.

The special aura surrounding command is especially visible in the dearth of analyses of command process, and of biographies of chiefs of staff, despite their vital role in implementing commanders' broad visions—and sometimes a good deal more. While historians have tended to focus on prominent individuals' actions and decisions, and to trace broad, clear formats and patterns, there has been some refocusing under the New Military History toward what were once seen as minutiae, and on the activities of what David Mechanic deemed "lower participants," even if under the implicit assumption that the latter, while interesting, are of minor import. While that sorting makes explanation and attribution of cause much easier, it is not clear whether it is more valid.

Beyond such questions, in military history, as in other variants of the discipline, considering the smoothing and fairing of details to make them fit patterns, assumptions, and conclusions takes on a different flavor when viewed from the perspective of nonlinearity. Close scrutiny of the command process suggests that the orchestration of operations by commanders in war lies far from being a close analogy to conducting a symphony orchestra. Some fragments of that intricate interplay have been captured for history in the form of messages, and entries in logs and journals, on maps and worksheets, although that has been less and less the case as advances in electronic communications and the increased pace of operations in mechanized warfare and secrecy led to decreasing use of written modes. At a further order of uncertainty, there is no way to be sure of how much such evidence aligns with reality.

In war, as noted at the outset, the preparing of such documentation in the press of operations has been shaped, to varying degrees, by dislocation, fear, stress, and fatigue. Outright destruction may not be evident to analysts or historians other than by their deducing a gap or discontinuity in the documentary flow. Since there will rarely be details on friction, anger, aberrant behavior, panic, and so on, trying to portray command-staff functioning in military history lies closer metaphorically to gross anatomy than to cellular physiology. Usually, but not always, that very complex "metabolism" is dealt with by allusion, or as a vague "given," under an assumption of some degree of rationality and order. On the surface of it, staff and command structures depicted in organization charts appear starkly rational, as do descriptions of staff duties and functions in manuals, regulations, orders, and after-action reports. That allows no direct view of what was in anyone's mind in a given operation, let alone what was happening, or whether leaders envisioned a gestalt, or tried to impose one, as opposed to reacting intuitively to confusing data.

In spite of that, the mystique of military leadership based on a mythos of special ability persists as a vision of generals or admirals at the center of a complex information web, recognizing and responding to the flow of data,[27] and striving to impose order on disorder. In spite of the value placed on commanders' visions, some generals who suffered major defeats and frustrations, like Hooker, Lee, Haig, Ludendorff, and Eaker, visualized patterns that proved to be well out of phase with reality but continued to try to implement them vigorously in the face of that volatility. On the other hand, some victors, including Schley, Yamashita, and Dowding, "winged it" in very fluid situations. While in battle, both sides grope for order in wild uncertainty and try to force a pattern on what appears to be chaos; that basic dynamic is manifest in many human activities, from the arts, science, law, medicine, business, and sports to criminality and war. In all of those fields, a full grasp of all details, even vital ones, often lies out of reach of the actors at the time, rendering the impulse to influence events a form of gambling.

The crafting of history to describe and judge such actions and events is based on the presumption that what happened can be seen more clearly in the course of reconstructing events, and that historians are better able to evaluate what occurred and whether what was done should have been done, or if there were better alternatives. The special challenge that complexity theory presents to that line of logic is the question that it raises of whether even the most laborious examination of the evidence—or the evidence itself—allows sufficient reconstruction of a fragment of the past. Although von Clausewitz's dictum that war is the "province of chance" is well known, its implicit premise, that wars are collisions of blind forces and more inchoate than the rationalized accounts of them suggest, is not widely accepted.[28] Most military history and memoirs have been aligned with templates of order, however coherent actual operations were, which is in keeping with what commanders throughout history, even such artful practitioners of evasion as Fabius Maximus, du Guesclin, and Giap tried to do. Interestingly, Giap encountered serious problems on at least two occasions when he tried to move from ambiguous diffusion to linear formats.[29] Such reflexive ordering has also been widely visible in military practice and operations, as in the naval practice of "shaking down," and the emphasis on linearity in ceremonies, and especially in rites of passage. That sometimes proved counterproductive and out of phase with the nonlinear milieu of war, but on other occasions, it yielded clear benefits, as in the upgrading of medical services in the American Civil War,[30] and the refinements of infantry tactics in the U.S. Army in the Korean War.[31]

Are such realignments and recasting of doctrines, models, and metaphors in the military-naval milieu analogous to the "paradigm shifts" adumbrated by Thomas Kuhn in *The Structure of Scientific Revolutions*? Or are they merely points on a gradient of chaos-complexity? The overall pattern of successes and failures suggests a need for some caution when relying on previous cases and linear models.[32] Once again, considering nonlinearity leads to divergence and variety, and denies surety about optimal conditions and the framing of neat categories. In a parallel vein, the history of both warfare and science are sprinkled with examples of amateurs outperforming seasoned professionals.[33] Both the paradigm shift and gradient models underline the lack of clear vectors in history pointing to the efficacy of specific types of doctrine, tactics, or strategy. For example, linear arrays and grinding tactics in the American Civil, Russo-Japanese, First and Second World, Korean, and Iran-Iraq wars were widely seen as wrongheaded,[34] while phased frontal efforts yielded success in such World War I battles as the Brusilov offensive of 1916 and the Australian-Canadian-British drive on August 8, 1918, and in World War II at El Alamein, in many amphibious landings, and in the crossing of the northern Rhine. In the most abstract sense, such highly structured tactics and fluid schema, like B.H. Liddell Hart's "indirect approach" and recent maneuver warfare schema, are extremes along a balance beam in which the risk of incurring some degree of chaos is weighed against accepting some degree of it in devising a way to foist complexity and uncertainty upon an enemy.

The military concept of the "fog of war" conveys the ambiguity that commanders and staffs grapple with, and the fact that they cannot see the whole matrix on which they plan and act. As noted in respect to command-and-control, technology has not provided as much of an advantage as many hoped, partly because of the fact that the reduction of the complexity of some tasks was offset by an increase in others,[35] in keeping with W. Brian Arthur's view that complexity has been the product of adding increasingly complicated subsystems to relatively simple systems to help them overcome limits, deal with anomalies, or "adapt to a world itself more complex."[36] In the broadest terms, that has been visible for at least four millennia in the evolving of military systems, when they are seen as analogical to machines, and composed of articulated parts. The forming of military forces into patterned formations is at least that old, whereas the symmetry of such formations, like a spider web, has always become tangled when committed to battle, and sometimes dramatically so, like a web's springing suddenly into a wildly different shape when a single anchoring strand is cut. Awareness of such susceptibility of complex systems to the tyranny of little events has been visible in popular culture, in the proverbial story "for want of a nail." A corollary in military theoretics is von Clausewitz's oft-quoted aphorism that in warfare, all things are simple, but the simplest thing is terribly difficult.[37]

The recurrent dissonance between the impulse to ordering and the complexities of warfare has long confounded analysis. Not all the ramifications have been explored, assuming that they could be, nor have those that have been examined been traced out to their full implications, such as Lewis Richardson's observation that an increase in complexity, in the form of the numbers of combatant nations involved in wars, has tended to reduce their bloodiness.[38] No one trying to impose order on warfare, either in operations or in the crafting of history, can escape the law of location, nor will those trying to do so from a nonlinear perspective find it easy to abandon the mythos of patterns lying beneath the turbulent surface of things in the waging of war. Of course, neither military history nor chaos-complexity theoretics and research points to a wholly random tangle in the dynamic of battle, but they do raise the probability that both order *in* and the ordering *of* the complex mechanics of warfare are rarer and less certain in dimension and form than normally envisioned. Coherence has been most visible in war in such activities as supply and personnel flows, procurement, medical evacuation, and artillery fire control, whereas wilder random oscillations have been more apparent in battle dynamics. Artful phasing of elements by commanders is depicted in many military histories, even though it is often very hard to determine how much the outcome of a specific battle was really a product of design and orchestration beyond the result of assemblage and launching of forces along general vectors.

There has certainly been a wide variety relative to coherence in warfare, from the rigorous, detailed planning, delineation, and phasing in amphibious and combined arms operations, and the meticulous style of Monash, Montgomery, and Ridgway, to the hell-for-leather tactics of Montrose, Stonewall Jackson, Nathan Bedford Forrest, and Heinz Guderian. Guderian, like Rommel, was not only reacting against abstract academic concepts like the German army's World War I "surface-and-gap" doctrine, but recognized, along with his peers, that positional warfare had chronically yielded high losses, and that the *Reichswehr,* tiny by the terms of the Versailles Treaty, could not afford to endure even a little of that. Like late-seventeenth- and eighteenth-century generals, Guderian and other advocates of mechanized warfare, on the ground and in the air, hoped that increasing speed and maneuverability, by allowing swift concentration and dispersion, would offset superior enemy firepower.

Like all reformers, they did not shed everything from the past, since as they broke many ties with it, their education and socialization set limits on their departure from foundations. Such tempering of vision by experience and habituation, and melding of continuity with inspiration, can be seen throughout military history. In the U.S. and British navies between the World Wars, aircraft carriers were subordinated to assumptions of a decisive clash between battleship fleets, and the searing trauma that was inflicted on them by submarines in World War I was discounted as an anomaly, unlikely to be repeated. On a parallel track, air power and tank advocates, who saw themselves as avant-garde, did not break wholly free of the gestalt of massed formations of aircraft and tanks. When war came, armor commanders altered their tactics in the face of improved defenses, somewhat more willing, or at least forced by the intricacies of terrain and ground war, to diversify than the air commanders, who again and again sent forth ordered, massed bomber formations from 1936 to 1945 as, in spite of heavy losses over Spain, China, Britain, Malta, and Germany, they dismissed or overlooked the potential that air power offered for asymmetrical diffusion and vectoring.

During World War II, the use of air power in smaller, less predictable increments and varying formats was infrequent, marginal, and uneven in effect. As deployment of fighter escorts was tied to large aerial parade-ground formations, the linearity and order of the previous century prevailed, blocking a perspective that chaos-complexity research has made more apparent, namely, that apparent disorder, like camouflage painting, might mask an inner pattern of "statistical regularities within chaos" that might be discerned "provided that chaos is used as a probe."[39] Only in the final phases of the air campaign against Germany and Japan did U.S. air leaders abandon daylight bombing en masse, and they returned to it in Korea, Vietnam, and the Gulf War.

However obvious it is that relying on symmetrical tactical formats offers an opponent a clearer view of resources and intentions, since repetition and patterning offer predictability, in military history one finds relatively few commanders and planners at the head of organized forces abandoning order for randomness, compared with guerrillas, insurgents, and tribal warriors. There are, however, many cases of the former resorting to a novel format, such as Hannibal's kaleidoscope of tactical gestalts in the Second Punic War and the depredations of the Special Air Service in the Western desert in 1940–41. Such instances of a redefinition of order suggest that there might be some value in nonlinear theoretics raising sensitivity to neglected or alternative configurations and potentials, and at least offering an antidote to false certainty, and an antidote to smugness arising from seeking out and imposing order for its own sake.

NOTES

1. E.g., see E.S. Quade, "Pitfalls in Systems Analysis," in his *Analysis for Military Decisions* (Santa Monica: Rand Corporation, 1964) [R-387-PR], pp. 300–316.

2. Quoted in P.M. Skirdo, *The People, the Army and the Commander* (Washington, D.C.: U.S. Government Printing Office, n.d., c. 1978), p. 54.

3. For clinical perspectives on coping under stress, see Richard Lazarus, *Psychological Stress and the Coping Process* (New York: McGraw-Hill, 1966), p. 208; and D.R. Davies & G.S. Tune, *Human Vigilance Performance* (New York: American Elsevier, 1969).

4. Jonathan S.B. Evans, "Psychological Pitfalls in Forecasting," *Futures*, 14:4 (August 1982), pp. 258–259.

5. Hence the view of warfare as an "unstoppable engine," as examined by Daniel Pick, in *War Machine: The Rationalisation of Slaughter in the Modern Age* (New Haven: Yale University Press, 1993), p. 11.

6. Contrasting views are R.E. Stivers, "The Mystique of Command Presence," *U.S. Naval Institute Proceedings*, 94:8 (August 1968), pp. 27–33; and Edward K. Jeffer, "Generalizing: The Mystique of High Command," *Army*, 43:6 (June 1993), pp. 43–46.

7. John Keegan, *The Face of Battle* (New York: Viking Press, 1976), p. 173.

8. E.g., the flight from Paris induced by the Germans' long-range naval gun in 1918; from Chapei in 1932 under Japanese air attack; the refugees that jammed roads in Belgium and France in 1940; the flight from Hungary in 1956, from North Vietnam in 1954, and from South Vietnam in 1975 onward, and from Kuwait in 1990. In U.S. history, cases include the flight of Tories, from various Indian depredations, the mass movement of slaves and freedmen in the Civil War, and of millions from urban centers, in the Spanish-American War in 1898, as the result of Orson Welles's "War of the Worlds" broadcast in 1938, and the Cuban Missile Crisis in 1962.

9. See Robert L. Helmbold, *Decision in Battle: Breakpoint Hypotheses and Engagement Termination Data* (Santa Monica: Rand Corporation, 1971), p. 12; and James Taylor, "Recent Developments in the Lanchester Theory of Combat," in K.B. Haley, ed., *OR 78* (Amsterdam: North Holland Publishing Co., 1979), p. 773.

10. Examples are Raymond Henri et al., *The U.S. Marines in Iwo Jima* (Washington, D.C.: Infantry Journal Press, 1945), p. 46; and Alan Moorehead, *Gallipoli* (New York: Harper & Row, 1956), pp. 114–116.

11. Quoted in Sir Bernard Lovell, *P.M.S. Blackett: A Biographical Memoir* (London: Royal Society, 1976), p. 59.

12. E.g., Edward B. Hamley, *The Operations of War Explained and Illustrated* (Edinburgh: William Blackwood, 1923), and J.A. Stockfish, *Models, Data and War: A Critique of the Study of Conventional Forces* (Santa Monica: Rand Corporation, 1975) [R–1526–PR], p. 130.

13. For an example of the tabular arrays common in Soviet "military-historical science," see V.I. Chuikov, *The End of the Third Reich* (Moscow: Progress Publishers, 1978), p. 173.

14. Jomini did recognize the limits of mensurability of military phenomena, as in his concluding that it was "difficult to lay down rules" in the analysis of amphibious operations, cf. Baron Antoine Jomini, *The Art of War*, trans. G.H. Mendell & W.P. Craighill (Westport: Greenwood Press, 1971; orig. publ. 1862), p. 229.

15. Thomas Buell, *Master of Sea Power* (Boston: Little Brown, 1980), p. 199.

16. E.g., see Madan M. Gupta & Ellie Sanchez, *Fuzzy Information and Decision Processes* (Amsterdam: North Holland Publishing Co., 1982).

17. R.L. Gregory, *The Intelligent Eye* (New York: McGraw-Hill, 1970), p. 166.

18. Kent Roberts Greenfield, Robert R. Ralmer & Bell I. Wiley, *The Organization of Ground Combat Troops*, Vol. 1 (Washington, D.C.: Army Ground Forces Historical Division, 1947), p. 236.

19. Quoted in Gordon Craig, *The Politics of the Prussian Army* (New York: Oxford University Press, 1964), p. 382.

20. E.g., the U.S. Army list includes objective, offensive, simplicity, unity of command, mass/concentration, economy of force, maneuver, mobility, surprise, and security; for a recent analysis of the concept, see John I. Alger, *The Quest for Victory: The History of the Principles of War* (Westport: Greenwood Press, 1982).

21. Trevor N. DuPuy, "Perceptions of the Next War," *Armed Forces Journal International*, 118:10 (May 1980), pp. 40, 50, and 54.

22. B.H. Liddell Hart, *Strategy* (New York: Praeger, 1965).

23. John Wheldon, *Machine Age Armies* (London: Abelard-Schuman, 1972), p. 134.

24. B.H. Liddell Hart, ed., *The Rommel Papers* (New York: Harcourt, Brace & Co.), pp. 507–524.

25. J.F.C. Fuller, *Machine Warfare: An Inquiry into the Influence of Mechanics on the Art of War* (Washington, D.C.: Infantry Journal Press, 1943), p. 61.

26. David Hackworth, "The Warrior," *Houston Chronicle*, September 20, 1981, Sec. 3, p. 23.

27. E.g., see William H. McNeill, "Mythistory, or Truth, Myth, History, and Historians," *American Historical Review*, 91:1 (February 1986), p. 2.

28. Perspectives on Clausewitz in respect to complexity theory are changing, e.g., see Christopher Bassford, *Clausewitz in English: The Reception of Clausewitz in Britain and America* (Oxford: Oxford University Press, 1994).

29. See Vo Nguyen Giap, *The Military Art of People's War*, ed. Russell Stetler (New York: Monthly Review Press, 1970), esp. pp. 100–106.

30. *Facts About the American Civil War* (Washington, D.C.: Civil War Centennial Commission, 1960), p. 10.

31. Robert Doughty, *The Evolution of the U.S. Army Tactical Doctrine, 1946–1976* (Fort Leavenworth: Combat Studies Institute, 1979), pp. 8–9.

32. See Judith Clarke, "First Punches: Review of Analogies at War: Korea, Munich, Dien Bien Phu and the Vietnam Decisions of 1965," *Far Eastern Economic Review*, December 3, 1992, p. 3.

33. E.g., Arminius; Spartacus; Hereward, the Wake; Joan of Arc; Cromwell; Nathaneal Greene; T.E. Lawrence; Hitler; and Vo Nguyen Giap.

34. E.g., see John J. Mahon, "Civil War Infantry Tactics," *Military Affairs*, 25:3 (Summer 1961), pp. 57–68.

35. For a discussion of differences between technical and task complexity, see E.W. Paxson, M.G. Weiner & R.A. Wise, *Interactions Between Tactics and Technology in Ground Warfare* (Santa Monica: Rand Corporation, 1979) [R-2377-ARPA], p. 36.

36. W. Brian Arthur, "Why Do Things Become More Complex?" *Scientific American*, 268:5 (May 1993), p. 144.

37. Hence Herman Kahn's concern that the unstable nuclear balance might yield major reactions when affected by small changes, cf. *On Escalation* (New York: Frederick Praeger, 1965), p. 144.

38. See David Wilkinson, *Deadly Quarrels: Lewis F. Richardson and the Statistical Study of War* (Berkeley: University of California Press, 1980), p. 118.

39. Heinz R. Pagels, *The Dreams of Reason: The Computer and the Rise of Science of Complexity* (New York: Simon & Schuster, 1988), p. 83.

7

Matching Frequencies:

Complexity and Creativity

Interest in creativity dynamics rose sharply in the 1950s and 1960s in academe, the business world, and the intelligence community, then subsided in those realms while gaining popularity in the "counterculture" of the 1960s and early 1970s. Unlike systems theories which had practical applications and could be expressed mathematically, creativity remained speculative, hypothetical, and ambiguous. Until the 1980s, whether due to abstruse concepts and language, or to the nonlinearity of the creative process itself, there was far less interest in such matters in American defense circles than in the U.S. intelligence community,[1] or the Soviet armed forces.[2]

Before attempting to examine links between creativity, complexity, and military matters, it is necessary to confront the dilemmas of definition. Dictionaries offer little help beyond referring to creativity as that which is "novel" or "new," and the diverse terms, effects, and concepts in each domain further complicate the comparing of phenomena. Yet there are widely shared if fuzzy views of creativity as innovation, the producing of fresh devices, concepts, or patterns by slightly altering existing forms; and as invention, the generating of wholly new forms. The essence of the former is caught the expression "Now, why didn't I think of that?" and the essence of the latter in "I never would have thought of that." Popular culture images include such stereotypes as Archimedes' exclamation "Eureka!" and cartoonists' light bulbs that suggest inspiration. Revolutionary inventions or concepts of great magnitude, referred to hyperbolically as "breakthroughs," "quantum leaps," or "paradigm shifts," are closer to the ultimate tribute to originality: "How could anyone have thought of that?"

Theorists have constructed models of creativity as a process running from problem recognition or definition through the generating of solutions, including mixing and matching, insight and inspiration, to selection, and sometimes on to a stage of development and application. Such attempts to reduce a highly "fuzzy" and semiconscious dynamic to concrete formats and rubrics are confounded by the fact that many creative phenomena occur outside such sequencing. Some inspirations emerge suddenly and very well formed at the conscious level with no previous definition of a problem, and many solutions that were generated in a focused and structured response to clearly recognized quandaries proved inadequate, hence the aphorism that a writer's best friend is the wastebasket. The conceptual and semantic slushiness surrounding discussions of creativity has impaired both consideration and gaining of acceptance of its utility, in spite of attempts to demystify, enhance, and use it by such analysts as Frank Barron and Edward de Bono.

As with complexity, creativity theory takes on a substantially different hue when considered in the context of military operations. It is obvious that the costs of military leaders falling in love with their own ideas or rejecting novel stratagems can be much higher than in such activities as commerce, science, and politics. Here, the drama, scale, and consequence of warfare distorts perspective. Even though both orthodoxy and creativity have each yielded successes and failures in battle, the image of military professionals, from generals and admirals to NCOs, being hostile to novelty and insight is nearly universal. That is to a great extent owing to the fact that creativity is a process that begins with individual human thought and behavior, and has its roots in the sub- or unconscious, as reflected in the popular "aha" and "Eureka" images. Again, inspiration and insight are not tractable to routinization and sequencing, that makes the managing of it, which really means the managing of people with creative capacity, very difficult. Not only is it hard to select people who will be creative, but some creative people are stylistically given to being difficult (the so-called prima donna syndrome). The "dialogue between harmony and invention" has often been shrill. History abounds with cases of inventors being frustrated, or meeting unpleasant ends, and prophets have been routinely ignored since Homer created the mythic image of prophets without honor in their own land in the character of the Trojan princess Cassandra in the *Iliad.*

In *The Psychology of Military Incompetence,* Norman Dixon asked why there was such a misalignment between warfare, the most chaotic, tumultuous of human activities in scale and consequence, and the rigid militaristic attitudes and reflexes identified by the stereotype of the "military mind." Although many military leaders have been flexible and shrewd, such cases are generally seen as exceptions to a rule. There are, of course, strong tides running against creativity in armed forces.

At the first level are mechanisms of recruitment and selection into institutions that proclaim themselves to be bastions of linear orderliness and formalistic hierarchical authority. Beyond that lie other screens and thresholds, including rigorous physical conditioning, the stressing of simplicity as a virtue, and obsessional fixation on details in training and socialization. All of those are aimed at toughening and developing will; they have also served to screen out or suppress creativity, both as process and trait. The effects are often no different on conscripts than on professionals.

Beyond entry-level training and rituals, those traits and reflexes are also reinforced at each echelon. From the intensely conformist subcultures of small units, ships, and aircrews to the upper tiers of bureaucracy, both parochial and bureaucratic forces tend to stifle initiative and individuality. The arguments can be made that it is necessary to maintain discipline and assure compliance, and that the effect has not been universally suppressive of novelty. In spite of such exceptions, from Hannibal and Caesar to Napoleon and Stonewall Jackson, a tendency to suppress such uniqueness has been recognized by practitioners and bystanders. General John Wood argued that "individual hazing has no place in the formation of a true military character," since armies needed "sensitive, intense natures . . . as well as the thicker-skinned, hard-boiled types,"[3] and the commander of the Australian Corps in World War I, Sir John Monash, advised a friend, "Don't let your son go into the Regular Army. It's the narrowest of all training."[4]

Some of the resistance to novelty in military and naval systems arises from the fact that professionals are drawn to the predictability, order, and security that military institutions provide in peacetime, which helps to explain why, when various technologies offered a significant advantage, inertial reflexes led to rejection or a fumbled opportunity.[5] While they have dominated the setting of forms and standards of armed forces, they are not the only element in society predisposed to linearity and regularity. In the United States, many males see creativity, especially such linked traits as sensitivity and tolerance for ambiguity as feminine, effete, slovenly, or linked to quirkiness or mental instability.[6] In spite of that, many instances of novelty in military history stand in tension with the pacific views of many creativity theorists and enthusiasts, a paradox recognized by students of creative process.[7] The artful crafting of weapons and torture devices, in blending viciousness with innovation, embodies both a fusion of creativity, and chaos, since torture inflicts chaos upon individuals in the same sense that war does upon organizations and cultures.[8] The use of the word *artful* may seem semantically perverse here, like the "art of war" and "operational art" as used by military professionals, since war and art lie at opposite poles of destruction and creation. Nevertheless, generalship has often been described as a kind of performing art, since, like artists, military commanders have often sought to create new forms, albeit on vastly disparate matrices and with very different materials. From the standpoint of complexity, creativity both in the arts and in warfare aligns along a "chaotic trajectory" within a bounded region, and the perception of creativity and its obverse, stagnation, are determined by whether the curve is divergent or reiterative. Both military history and the "alternative futures" genre of science fiction and fantasy are replete with scenarios in which major outcomes hinged on small forces that moved unfolding events from a minor to a major scale of reference.[9] Another gross congruence of warfare and art is that each is suffused with deep passion as artists and warriors strive to impose order, and in each, style is seen as crucial.[10]

There are obvious dissimilarities. Stereotypically, artists are seen as striking poses of nonconformity and disdaining bourgeois standards of dress and deportment, whereas military culture appears to have strong, clear lines of form and process that maintain form and coherence in the turmoil of war. In spite of such stereotypes, military institutions in peace and war are not wholly linear and rigid, any more than the world of art is total bedlam.[11] In the course of modern wars, armed forces have tended to move toward more individualistic and casual behavior and dress, partly as the product of adapting to conditions, such as muting colors and dulling the shine on metal, and some of it has been flamboyant idiosyncracy, a casting off of orthodox forms and procedures. Thus wars, in the modern era at least, have carried within them the seeds of social unrest, revolution, and mutiny.

Not only military habits, practices, and attitudes have shaped the proverbial canvas on which commanders have attempted to render their concepts in the form of tactics, for that matrix has never been wholly blank at the outset. As noted previously, until the mid-nineteenth century, heads of state often controlled military operations in person on the battlefields, while since then, Napoleon III, Lincoln, McKinley, Hitler, Churchill, Stalin, and Lyndon Johnson, among others, did so over long distances through the medium of communications networks. Aside from such exotic cases as the Viking raiders, Attila, Genghis Khan, and Tamerlane, military leaders have usually been constrained in conducting war by broad directives from higher authority, regulations, policies, domestic and foreign public opinion, laws, values, ethics, and customs. Other strictures include the emotions, anxieties, and fatigue that affect commanders and staffs as they practice their craft amid great turbulence; in the relentless glare of history's spotlight they have no counterpart in other human activities. Of course, military professionals in peacetime can only vaguely sense those cross-strains or anticipate the future matrices on which they might practice their art, or, indeed, if they will ever do so at all. If that comes to pass, they know that some of what they will see will not be real, as they work like painters or sculptors in very dim light—or darkness. With that irreducible gap between expectations and reality in view, Captain Wylie cautioned, "We cannot with certainty predict the complete pattern of the war for which we prepare ourselves."[12] In similar frustration, during the Vietnam War, an American operational researcher lamented the "lack of combat data, at least of the kind that can be analyzed in a systematic fashion,"[13] echoing a complaint made a generation earlier of a "total" lack of "adequate means of front-line demarcation in northwest Europe in World War II."[14] Such inadequacies of human perception and control in dealing with complexity are in keeping with R.V. Jones's observation that "witnesses were usually right when they said something had happened at a particular place, although they could be wildly wrong about what had happened."[15]

In spite of such perceptual limits, many attempts have been made to quantify warfare and render portraits of it closer in style to realism than impressionism. While the best-known bellometrician is Jomini, there were many others. Since the Renaissance, quantification has been brought to bear on battle dynamics repeatedly (e.g., the near-cult of "angle of attack" theory in the late eighteenth century). A good deal of it was driven by the applications of algebra, geometry, trigonometry, and calculus in siegecraft, ballistics, and navigation, and variants of it have appeared periodically in the history of modern warfare, usually in response to changes and challenges produced by new technologies. A major resurgence appeared in the late 1930s in the form of "operational research" in Britain, in which teams of scientists from different disciplines addressed first air defense networking and later a host of problems in land, sea, and air warfare. Reliance on such numerical analysis increased even more dramatically after World War II as electronic digital computers, nuclear weapons, and aerospace warfare planning converged with various concepts of cybernetics and systems analysis. Logics and systems based on precise quantification came to dominate doctrine and "hardware" alike in the arcane field of thermonuclearetics, and then in planning, budgeting, and policy formulation and analysis.

The trend toward rigorous quantification accelerated in the 1950s, suppressing but not wholly eclipsing consideration of ambiguity and complexity in warfare. The introduction of computers and systems analysis virtually revolutionized the Pentagon in the early 1960s as Robert S. McNamara, a former president of Ford Motor Co., and his "whiz kids" entered the scene. The concept of the art of war gave way to a business of war as their concepts eventually permeated governmental budgeting and planning at all levels, and foreign military systems as well. The military professionals were initially dismayed but moved quickly to master the new methods as the Vietnam War massively warped the U.S. defense structure. The confounding of those expectations and methods in that conflict led to discrediting of numerically based analysis of combat processes, and the study of military history was rejuvenated to gain perspective on precedent and method as well as such "squishy" aspects of war as chance, genius, leadership, and creativity. Although some leaders in the past were seen as having been especially able to deal with the confusions of battle, such appraisals overlooked the question of how well someone else would have done in the same situation—and the fact that several "Great Captains," including Hannibal, Napoleon, and Rommel, ultimately lost.

As Western analysts of command prowess framed a concept of "operational art," drawn from Russian military doctrine, judgment of performance remained more a matter of impression and preference than a product of methodologically consistent analyses. With no firm template for effectively measuring or predicting such capacity, two vague sets of assumptions can be found at work in military history and the training of commanders. One is a kind of dim faith that anyone thrust forward by fate to command in war out of a cohort of elites is likely to have a reasonable amount of command ability. The other is a contradictory view of excellence in command being innate, and encompassing an intuition analogous to artistic giftedness. Each of those concepts conforms with Frank Barron's postulate that ". . . the creative individual is someone who has an exceptionally strong need to find order where none appears, and who as a result of his own abilities and personal experience honors the apparently unclassifiable with his consecrated attention."[16] That applies as much as to Napoleon, Nelson, and Rommel as it does to Freud, Edison, and Picasso.

In other respects, however, the task of commanders and staffs stands closer to science than art. As professionals, they conform closely to the following description of scientists as "deeply conservative . . . [and] indoctrinated into a paradigm," that is, sharing as a central, mutual, and accepted pattern of thought a kind of superdoctrine. As they try to solve "puzzles," they encounter "anomalies." Although these are often ignored, "if they accumulate . . . [they] may trigger a . . . paradigm shift," proponents of which are "often young or new to the field, that is, not fully indoctrinated. . . ."[17]

Whatever parallels there may be between them and other categories of leaders, military and naval commanders confront the most intense stresses and the gravest consequences if they fail, as they strive to maintain balance and purpose within a maelstrom of violence. Those bewildering conditions place a premium both on the abilities to impose order and to accept imprecision and ambiguity, and if possible, to harness them. Edward de Bono caught the essence of that paradox in hypothesizing that it was "blurry-brained creative people" who "established new general ideas," and that "sharp brains" were "essential but only for refining, developing and using the ideas thrown up by blurry-brained thinking,"[18] as did Yosimichi Yamashita in observing that "to be part of something wonderful happening in the minds of the people within a large organization, the manager must learn to be stupid."[19]

However valuable such traits are at higher echelons, armed forces require NCOs, technicians, and junior officers at the level of execution who take "a real delight in detail and in the exercise of personal supervision over every branch of work." Yet the training of officers to do that may produce leaders like those admirals in the nineteenth-century Royal Navy—and in other armed services at other times—who are "maximus in minimis—very great in small things."[20] At ascending levels of command, leaders and planners deal with a proportionally greater array of events and contingencies, which sharply increase in number and intensity in war. At each level of command, leaders face a wide array of future possibilities, immediate and long range, that fan out across the wide-flared lip of a figurative bell-shaped curve whose open end is aimed like a radar dish at the future. At lower tiers of command, the commanders' range of view forward into time is smaller, and the lateral range of contingencies narrower, since tactical leaders engaged in battle or in immediate anticipation of it cannot focus on more than a few contingencies close at hand in time and space. Strategists and high commands atop their larger "bells" survey a much broader scene that stretches further off in time and space, out toward the farthest reaches of the spreading base of the bell, where there is less clarity of detail, but much greater complexity.

In long-range planning in peace and war, commanders and staffs trace lines and vectors toward the various distant points in the future. Doctrines, plans, and policies may serve to buffer the anxiety born of such uncertainties, but since those are all different forms of assumption, they may not be much more than talismans. At both tactical and strategic levels, there is no sure way to narrow the gap between what is expected or hoped for and what actually happens. There are rough analogies to that dilemma in business, politics, and economics, but in war and accident, the unanticipated or implausible, so frequently commented on by veterans and victims of war and catastrophe, occur on a much greater scale and more often. In war, the clashing of the opposing systems increases the occurrence of exotic events as the lips of each side's metaphorical bell are extended by the force of that collision into ranges of more extreme probabilities. In a military variant of Parkinson's law, those habituated to focusing closely on a few details across a narrow range ascend the rungs of the military hierarchy, passing into an environment in which traits required to function effectively are the reverse of those that led them to success at lower levels, as they follow an analogical progression from mechanical skills through science to art.

Some parallelism between the military art and painting and sculpture is evident in the subjective evaluation of the performance of commanders, including the awarding of the status of "Great Captains." Much military historical writing resembles criticism in the arts and music in tone, and in its relying on impressions of style rather than unambiguous criteria. There has been far less effort devoted to framing a list of "principles of command" to use as a template for scoring leaders' performance than has been aimed at identifying "principles of war" (i.e., operational characteristics). The relegating of staff officers to relative historical obscurity is matched by the lesser glory awarded to leaders who devised unique formats or stratagems based more on craftiness than force majeure, like Belisarius, Montrose, du Guesclin, and von Manstein.

Whatever criteria might be used to assay military performance, it is an area in which chaos and creativity overlap. That is also true in science, and novelty in both domains arises from leaping to the outer strands of an ever-diverging stochastic web, or to the outer portion of the flared lip of a bell-shaped curve. Like artists and scientists, military professionals' awareness of the lineage of their craft is uneven. Some serving officers have been motivated to supplement the sparse and hasty exposure to history in their early formal education, usually blurred by hazing at service academies, and by other distractions in officer training at civilian universities. Such attempts to broaden professional knowledge may be reinforced along the career path at higher war schools or in assignments at universities, but overall, career pressures allow few chances to develop a broad configurational sense, or to interpret. To return to history, in the military profession, as in the arts and sciences, awareness of or dependence on previous forms and patterns may be seen as evidence of orthodoxy and academicism, or a form of pedantry unrelated to actual talent and performance. Not surprisingly, the range of historical cases drawn on to serve as prime examples and common references is relatively limited.

While it is not readily apparent that command success has been related to depth of knowledge, it would seem that enlarging the trove from which analogies and models are drawn should increase potential combinations and therefore enhance the creative process. Has the frequent failure of plans and expectations based on specific historical cases been due to a general irrelevance of the past to the future, to a lack of awareness of more appropriate models, or to overdependence on a single model? Is it worthwhile to provide military professionals or civilians responsible for military affairs a broad knowledge of the history of warfare, so that, like artists, they can strive to avoid clichés and work toward a fresh style? Just as complexity theory calls into question the value of extrapolating from one instance to another, a broad awareness of military history offers some a sense of the influence of discontinuity and accident in war, but awareness of many models and analogies might also restrict moving beyond innovation toward invention or insight, as some have argued that the "ghost of Napoleon" did in the American Civil War and World War I. Military professionals' relying on history has sometimes been derided, as in the cliché that military and naval professionals prepare for the next war by planning for the last. In keeping with that, fixating on one or a few models may not only blunt appreciation of the high probability that the seemingly unlikely is likely to happen in war, but it may also offer a foe an advantage by pointing to likely courses of action, while adding to the astonishment of those fixating on a particular model.

The confounding of simplistic, linear models in the entanglements of war has, like anomalies in other branches of history, led some historians, biographers, and memoirists to try to smooth history into congruence with the actual outcome, with orthodoxy, or reasonable expectations. Both creativity and complexity theory raise the question of whether it is better to generate many contingencies, even if some seem bizarre and zany, rather than risk the rigidity arising from fixating closely on specific possibilities. That question is resolved in democracies, where concepts and doctrines, in peacetime at least, must pass political review, unless they are masked in the "black" budget. As those forces impinge on innovation, including service and vendor interests, lobbying, and political "horse-trading," the contingencies of battle and military professionals' "best professional judgement" are often pushed aside, until war forces a converging of creativity, complexity, and chaos. The result is a version of that syndrome in science identified by Bohm and Peat stemming from "an excessively rigid attachment to fixed 'programs' in the tacit infrastructure of consciousness . . . [which] primarily . . . prevents . . . creativity from acting."[21]

The warping of procurement, development, and preparation out of phase with the contingency of battle by such forces also adds to the jarring effects of making the transition from peace to war. Often, in that process, as in the arts, talent had prevailed over seniority and orthodoxy. The emergence of leaders from outside the established elites, such as Nathaneal Greene, U.S. Grant, Erwin Rommel, and Lawrence of Arabia, is in keeping with the observation that:[22]

> Leadership in a disaster . . . [is] directly related to [the] type of situation produced. . . . Formally designated leaders often fail . . . because their skills do not cover the types of situations . . . [and] individuals [whose] skills . . . may be more effective in taking the leadership role . . . [and] may normally occupy a minor or insignificant role in the formal social structure. . . .

Despite all the dramatic cases of confounded expectations that can be found in history, in the corporate memory of institutions, and in folklore, there has, until relatively recently, been surprisingly little recognition in the shaping of formal policy and doctrine of the high probability that things will go wildly awry in war. Journalists and the general public are frequently astonished by revelations of policy failure, command error, and "friendly fire," even though they are common in war. Yet there are no clear norms, nor is the likelihood of such "glitches" weighted in theories of statecraft and war proportional to their chronic incidence, or to their potential consequences. Rather, in both policy analysis and history, judgments are usually seen to be founded on rational appraisals, based in turn on bargaining, linear trends, and variable models of balances and symmetries of power, in which options are derived from careful calculation of advantages and penalties.

Such rationalistic images and models lie far from the actual turbulent, unexpected, and sporadic events of wildly varying frequencies, amplitudes, and forms that leaders grapple with in war and crisis. Like canoeists trained on a quiet millpond to shoot rapids, when they actually enter the raging stream of crisis or war, and struggle to adjust to major state-changes, they abandon the sense of rhythm and reflexes developed in conditions of smooth order and find themselves far less concerned with steering a designated course than trying to stay upright. In that metaphorical transition, decisions and actions are the products of a mix of deliberate thought and reflex, the latter being the special kind of creativity that the Russian psychologists call "operational thinking," which lies closer to the performing arts than policy science. That, like part of the creative process, lies off the track of conscious thought, and, not surprisingly, has also often been linked to psychological aberrance, from preliterate societies' reverence for the seizure-prone to Schiller's view of creativity as "momentary passing madness."

Although many famous commanders have been seen as quirky and bizarre in demeanor and style, historians and journalists have tended to grant senior commanders a great deal of latitude in regard to eccentricity, in wartime and afterward. In modern times, however, generals and admirals have increasingly been expected to present a placid and controlled demeanor. While they have committed their lives to commit violence on a large scale if necessary, most have observed custom, law, edict, or ritualized forms of feudal honor, holding their warrant to wield destruction from the state in such a way that their applications of force fell short of wholesale mayhem, and was aligned with policy and purpose—even Sherman's March to the Sea was. In spite of the widespread view of military professionals as "trained killers," command in war has been seen by many warriors as the ultimate opportunity, within the metaphor of the art of war, to paint on a real canvas rather than to throw a tantrum. Such chances are very rare, since most military and naval professionals spend their careers, analogically, in sketching and reading about other painters. As a result, in the transition to war, many military leaders accustomed to harmonious long waves of peacetime routine have failed to cope with the uneven and massive fluctuations of combat, rather like high society portrait painters suddenly thrust into an advertising studio.

It is not surprising, then, that commanders' behavior resembles "artistic temperament." The idiosyncracies of such commanders as Nelson, Stonewall Jackson, Sherman, Grant, Montgomery, and Patton meshes with research on the emergence of visionary leaders in social systems under great stress,[23] and on links between creativity and neurosis.[24] Thus the metaphor of complexity, of water flowing from the glassy calm of a millpond through the segmented sworls of the millrace to the fully chaotic rapids below, not only symbolizes the state-changes from relative tranquility to raging chaos that military leaders endure in war, but states of mind as well. Awareness that such surges may occur does not necessarily grant a capacity to control those complex processes, or to adequately anticipate those tumultuous conditions. Whereas in art, critical judgment is based on the appraisal of a completed artifact, in war, it is very hard to tell just what that is. Consequently, military historians may concern themselves with peripheral traits and behaviors of commanders, analogous to the sequence of brush strokes, or conversations with visitors, and not set forth in a systematic way. That can be seen in even the most well ordered battles, like El Alamein, in which events got out of the precise control of the commanders, and accounts of which convey only a fragmentary sense of what exactly they did as events unfolded.

A sharper flavor of the inchoate can be found in eyewitnesses' accounts of actual combat rendered close to the event. War correspondent Robert Sherrod, who landed with the first wave at Tarawa, related how, with almost half of the three assault battalions killed or wounded, many survivors were reduced to a groggy state, since "their organization was ripped to pieces."[25] Yet that horror did not prevent the U.S. Marines from prevailing in that battle, or cause the higher command echelons to take full counsel of their fears. Just as such chaos and disorder in battle has not been fully apparent to those directing operations, it has also frequently been muted by authors who were striving to be clinical and objective, or to present impressions in the form of passing glimpses and anecdotes, and a sense of a underlying order and progression is conveyed in maps, graphs, and tables that give coherence to the glimpses of portions of a vast tapestry. Outcomes appear as aggregate products of large, indeterminate forces, and victory appears as the somewhat metaphysical offspring of a vast tangle of unseen and unmeasured events. It remains to be seen if advances in computer modeling and display technologies can significantly close the gap between portrayal and actual events in warfare, and aid in the identifying of such anomalies and details hitherto resistant to measurement and analysis, including nuances like anxiety and indiscipline.[26]

One amorphous aspect of war dynamics that touches on both complexity and creativity is the paradox that disorder, disintegration, and broken links between echelons have not always led to panic, nor to loss of fighting capacity. That was dramatically visible in the Wilderness campaign in the summer of 1864; as the Army of the Potomac moved south to fix and grind down the Army of Northern Virginia, there was no distinct front line. Veterans described the campaign, officially labeled as a series of distinctly separate battles, as ebbs and flows in a constant, undifferentiated clash of massive, formless mobs. Despite that functional disintegration, the high command's lack of a clear view of what was happening at several command levels and the widespread breakdown of organizational coherence did not blunt the fighting spirit of the bulk of the forces on either side.[27] Several such fluid, undirected "soldiers' battles" occurred in the twentieth century, including, in World War I, Verdun, Caporetto, and the German peace offensives of 1918, and in World War II, Bataan, many clashes on the Russian front, and the Battle of the Bulge.[28]

Those disintegrations of command in battle raise the crucial question of how much warfare at all levels, including battles that did not appear to be so clearly chaotic, has really aligned with designs and edicts of leaders, or with the description of them by journalists, or as they are rendered in official reports, memoirs, biographies, or histories. More essentially, are the portrayals of them by military historians closer or further from actuality than the accounts of participants, or even the ironic surrealism of *Catch-22* and *Good Soldier Schweik*? If that could be measured accurately, would the gaps between apparent and real control prove widely different from case to case, or something closer to a standard range of percentages or even a coefficient? Or is the dynamic of each battle such a variegated and unique product of elements, such as commanders' styles, doctrines, systems of training and discipline, plans, communications, technology, and logistics, so vast a collection of elements that it cannot be sorted out and properly arrayed for analysis?

There are very few fixed points on that murky and uneven landscape, nor even clear correlations between compliance and control and victory-and-defeat, or between discipline and battle performance. Both analysts and professionals, in trying to define military effectiveness and find methods and principles that might yield an advantage in the waging of war, have often decoupled tactical military performance as an art from ultimate outcome. Theorists have tended to focus on certain aspects of military operations abstracted from other elements. For example, after the Franco-Prussian War, in both Germany and France, special concern arose over morale, bonding, and unit cohesion, as it did in the U.S. Army after World War II, Korea, and the Vietnam War. A drawn-out debate began, arising from the ongoing fragmenting and dispersal of air, ground, and some forms of sea combat due to technical changes. Some argued for maximum delegation of decision-making to lower echelons, while others saw that advances in communication technologies might help high-level commanders exercise immediate control over vast forces arrayed over great distances. As noted earlier, the chronic confounding of high-level commanders' plans and concepts in the wars of Prussian ascendancy, 1864 to 1871, led the victors, perplexed by all that went awry in spite of their triumphs, to formulate the "law of the situation," in essence much like the "*doctrine a priori*" and "*doctrine des circonstances*" drawn up by then Colonel Charles de Gaulle half a century later.[29] Both, although elegantly framed, were tacit admissions that there were no clear rules nor effective instruments of prediction to aid those who tried to anticipate impending military operations.

In those debates over control and authority, little attention was paid to discipline per se, either by professionals or by military historians.[30] Although potentially vital in commanders' imposing their concepts and patterns, it was treated as a relative given. Indeed, by the mid-twentieth century, in the Western democracies, commanders' relying on coercion and formal discipline was widely seen as antithetical to effective leadership, a sign of incompetence or desperation. That changing view reflected both a liberalizing of values and the dispersion of warfare due to advances in technology, since the former eroded support for the bloody encounters of massed forces, and the latter required the dispersion of forces, which relied on intelligent, skilled operatives exercising initiative and judgment without close supervision. After World War II, that led to a moderating of leadership techniques, sometimes more in the armed forces of authoritarian states than in the Western democracies. Overall, discipline remained firmer at lower echelons than in the upper strata, although it could be argued that such a variance is visible throughout military history.[31] Indiscipline on the part of leaders was often seen as a sign of independence and high spirits, and creativity has been seen as linked to a disdain for norms, orthodoxy, and discipline, all barriers to reaching new forms and methods. The tolerance of those in higher authority toward such deviance has tended to be proportional to the deviant's success, but whether that is really a form of creativity is another matter. It is not clear if the chaos arising from easing discipline might work to cancel out resultant creativity, or how much firmness gains control at the cost of suppressing adaptive novelty.

Military creativity and optimum skill in the practice of "operational art" have often been defined in terms of commanders' surprising their foes. As Barton Whaley, Richard Betts, and others have pointed out, the occurrence of major surprise attacks in war has been increasing over the last century or so, as have elaborate schemes of deception. Beyond stratagems deliberately imposed by one combatant on another, the plans and expectations of winners and losers were frequently confounded by the unforeseen effects of combinations of forces and systems on their own side, and by the synergies produced by the full energizing of each side's forces that took place as they collided in war. (Such inadvertent surprise has not been scrutinized as closely as that resulting from deliberate design.) In spite of those trends, surprise and deception have generally received less emphasis in American tactical thought than the arraying and maneuvering of major elements of force in battle, an imbalance that may be considered in view of a chaos researcher's warning that "as time passes, unsolved problems within a given paradigm tend to accumulate and to lead to ever-increasing confusion and conflict."[32]

The evolving of communications technology also contributed to inadvertent surprise, although such systems were designed to augment higher echelons' capacity to direct war. Frequently, such systems proved less effective than designers or users hoped, while subordinates resisted, ignored, or evaded close supervision. As they grappled with the recurrent disproving in practice of many dicta, truisms, and nostrums, military professionals and analysts were unable to formulate what Stephen Hawking has deemed "a good theory," that is, the folding of descriptions of many cases of a relatively simple model into a predictive construct that conforms with subsequent observations.[33] In shaping doctrine and forming expectations, what seemed to be extraneous or marginal complexities were usually discarded or mentioned in passing. The sense of turbulence was also blunted by use of stark symbols, and unemotional language and style designed to cushion commanders from the extreme violence and emotionalism of war. Here again, while such insulating devices may allow better functioning, they can also contribute to the sense of shock and disorientation when battle is joined by creating too wide a gap between expectation and reality.

It has been argued, much in the spirit of artists' view of critics, that such matters can only be fairly judged by those who have experienced such jarring transitions in war, or cannot be judged at all. Yet in many cases, amateurs have beaten experienced professionals, just as in art the highest virtuosity is seen in transcending expectation by diverging unexpectedly from a sequence, and setting new forms and dimensions.

With that in view, Field Marshal Lord Slim warned young commanders against modeling themselves after others, holding that "imitations are never masterpieces."[34] In terms of complexity theory, military leaders' and staffs' creativity is expressed in their sensing, acting, and reacting as their "system is driven away from some equilibrium state and undergoes sudden transitions that, with increasing external stress, break the symmetry of the simple, least stressed state in more and more complicated ways."[35] That statement seems obvious enough, but many forces work against it, including bureaucratic procedures, concerns for safety in training, and adherence to civilian industrial and commercial standards, which widen the chasm between the realities of peace and war by creating a false sense of order and mensurability, or setting habits and reflexes that are based on irrelevant parameters.

The capacity of commanders to break away from the narrow track, discern fresh forms and patterns, and act with surety while imposing their will amid the turmoil of war has often been presented by biographers and historians as a kind of metaphysical orchestration. That is reflected in the lexicon of different military systems, for example, the Prussian term *Fingerspitzengefühl* (fingertip feel) a commander's intuitive sensing of the state of affairs in battle, *élan* and *cran,* French idioms for fighting spirit, and the Finnish term *sisu.* The semantics of command process, like those of creativity and complexity, are laced with such imprecision, allusion, and "fuzziness," and stand away from the assumptions of rational action and normative behavior implicit in bureaucratic regulations, laws, doctrines, and other organization rules. Most practitioners and analysts agree there is an intuitive element in the command process, but not if it is a "knack" that only some possess, or a skill that can be precisely identified, or cultivated.

In most armed forces, the capacity to reduce complexity to basics is seen as a primary command skill, hence the listing of "simplicity" as a "Principle of War," the American usage "cutting to the chase," Foch's aphorism *"d'quoi s'agit-il,"* and the German concept of *Schwerpunkt,* each conveying a sense that complexity should—and can—have an ordering template placed upon on it. In spite of all the visions of war as a deliberative process akin to engineering or architecture, many analyses of doctrine have shown that the aligning of military operations with expectations or plans has not been regular or predictable. In spite of systems and technologies devised to increase control, battles and campaigns in the last century and a half have often been disorderly clashes of forces, lacking precise tactical goals or formal schema, and far from the degree of control and response implicit in the frequently employed metaphors of orchestra conducting and managing an athletic team. While the battles of Gettysburg in 1863 and of Santiago in 1898 resulted from a confused collision of forces, the deliberate structuring of intricacy and complexity has, like other categories of stratagems, had mixed results. George Washington's intricate five-column march plan to envelop the British at Germantown in 1778 generated true chaos, but the very elaborate interweaving of schemes in Operation OVERLORD/NEPTUNE gained the Western Allies a successful lodgment in Normandy in 1944.

Cases of accident and blind collision producing unexpectedly favorable results, in broad, figurative terms, appear to have been the result of the superimposing of a randomly oscillating wave upon another, yielding more points of intersection or congruence than the overlaying of a long, smooth, linear wave on a choppy, random pattern. The implicit probabilism of that model, and of complexity theory as well, raises the question of how much of the clash of forces in war can really be observed, and whether, if advances in sensing such intricate processes bring that into clearer focus, a clearer and simpler view will emerge. Do such attempts at regularization, conceptual and practical, such as abstraction, reductionism, formulation, simplification, and military discipline, really impose order, or do they only create images of control that are in fact out of phase with the turbulence of warfare?

Many forces work against such efforts, and add to the obscuring of the dynamics of warfare. Such observers as Leo Tolstoi, Fletcher Pratt, Martha Gellhorn, James Jones, and Paul Fussell have pointed out how filtering, beyond censorship imposed for purely security reasons, kept the full savage violence and disorder of warfare out of public view. A similar masking and abstracting of those aspects of war can also be seen in military parlance, formats, and ceremonies. Those attempts to disguise and deny the manifold horrors of combat are understandable, since, over the last two centuries, commanders in modern war have been confronted with increasing levels of forceful fury in war, in scale, tempo, and technical complexity, along with the scrutiny of media and historians. Inasmuch as public attitudes have often diverged sharply from military professionals' attitudes and customs, it is not clear how much the latter's adherence to "feudal" or "tribal" forms, including training and rites of passage, has been a form of political reaction, a blind adherence to tradition, or the preservation of utilitarian forms that, as the product of evolution, are best suited to prepare those who must survive and function in war.

Reliance on such ordeals has a substantial suppressing effect on creative dynamics in military institutions. Of course, the use of rites of passage to award status, privilege, and the right to perform certain elite functions is pervasive beyond the military realm, from the circumcision groups of preliterate societies to British boarding schools. In the military context, however, the evolving of war from a relatively constrained ritual to its elaborate and massive modern variants has led to a special paradox. The near-bloodless threat displays of New Guinea tribal wars, the linear arrays of ancient, medieval, and modern armies and fleets all reflect a long, general trend, sometimes disrupted by perturbations like Hun and Viking onslaughts, and the "fighting Irish." Since that line began to waver in the mid-eighteenth century, the trend toward fluidity in modern battle dynamics has steadily increased the premium on individual initiative and adaptability. From the light-infantry "revolution" of the nineteenth century onward, the changing morphology of combat was frequently ignored, rejected, or unevenly matched to recruiting and training with, with resistance to adaptation on the part of elites usually overcome only by crisis, defeat, or frustration, and the rising premium on creativity and flexibility discounted.

That particular dialogue between harmony and invention has been uneven and unsymmetrical. Paradoxically, the creative process has often been linked to abandoning self-control and bringing suppressed emotions to the surface, and involvement in warfare has had that effect as well. That has not always led to a heightening of creativity, but a freezing or tensing often attributed to the emphasis on linearity and rigidity in military socialization. In another paradox, traits associated with innovative capacity, such as flamboyance, rashness, and unorthodoxy, are in tension with the sangfroid and self-control expected of senior commanders, but the match of demeanor with performance has been far from exact. In peacetime especially, conformity and appearance count for much in the selection of core elites, and there is relatively little concern for fresh approaches, inasmuch as the exact nature of future operations is usually as unpredictable as the performance of individuals. Not only is it difficult to generate and match solutions to problems on demand, or in a linear, scheduled way, but creativity often involves the producing and discarding of ideas, a process antithetical to the behaviors most favored in highly conformist systems in which rewards and promotions are controlled by the evaluation of subordinates by superiors.[36]

Another dilemma that appears in respect to the applying of creativity to military purposes is a variant of the "Law of the Situation," that is, a particular problem may be solvable by the increasing or diminishing of complexity—or both, and in several ways. Such ambiguity is not likely to be accepted by those searching for quick, clear, simple solutions. A related complication arises from the element of chance. In the history of science and technology, fortuitous accidents and apparent coincidences abound, including Alexander Graham Bell's crucial spilling of acid, the "contamination" of Alexander Fleming's petri dish, and Paul Ehrlich's housemaid's fortuitous heating of his slides. Many inventors and discoverers have recognized luck as a key element in their successes, but that is not so strong a tradition in the military realm. Yet some commanders and historians have been willing to see fate as a prime factor in shaping destiny. Von Clausewitz defined "chance" as one of his key "provinces" of war, and Napoleon asked of a prospective general: "Is he lucky?"

The metaphors of the commander as athletic coach and orchestra conductor suggest the imposing of precise, linear control on a relatively ordered sequence of events. In such analogies, accident and chance are recognized as having some minor influence, and intricacies and subtleties like role and process are often excluded; creativity and complexity are usually dealt with by allusion, anecdote or impressionism. The gap between the model of rational, deliberate thought processes and the convoluted, intermittent nature of consciousness, and especially the nonconscious actions of performing artists and commanders in war, led the Russian psychologist Boris Teplov, Pavlov's principal protégé, a former Red Army officer and a student of musical ability, to probe the concept of "operational thinking" in depth in his essays on command psychology.[37] While both complexity and creativity theory can be brought to bear on such problems, metaphorically speaking, they may not be much more than great rings of keys used to try locks in a vast and darkened mansion. They do throw into relief the immeasurability of warfare, and make the vision of firm, certain control over warfare or precise description of it seem very bold, if not arrogant. Although those clusters of theory and phenomenology do not offer the simplicity and clarity that military professionals value highly,[38] research in both domains does throw light on how far historical process, including warfare, lies from the coherent symmetry and patterning of most conceptual models framed over the last three centuries. As investigations in complexity, chaos, creativity, and "fuzziness" in the sciences open doors onto murky, tangled, uneven landscapes that seem so different from Newton's brightly lit formal garden, some of the initial sense of disorder and incomprehensibility may pass away. On the other hand, the discontinuities that have emerged so far may, like speed bumps on a darkly lit road, be forewarnings of absolute limits to human cognition, and more resemble corollaries of Heisenberg's Uncertainty Principle than the clear formulae and principles of Newton and Maxwell.

What are those delvings likely to produce in respect to the understanding of warfare as a phenomenon—or an aggregate of phenomena? Some are hopeful that a transition is under way to more subtle and muted forms of conflict.[39] Nevertheless, many people, otherwise gentle and peaceful, of high intelligence and noble purpose, who have engaged in war, or prepared for it, have used their creative powers to do violence. The lofty ideals and universalist visions of scientists have not prevented the open boundaries of "pure" science from being raided, and elegant and abstract theories being harnessed to such grim purposes as the development of chemical, biological, and nuclear weapons, rockets, and napalm. The history of theoretical physics drives home the unsettling truism that what appears "pure" may ultimately have significant military utility, and that the most elegant and refined forms of human creative vision can be bent to the impulse to do deliberate violence.

Such applications—or perversions—of the creative impulse have generated orders of complexity and chaos in the military realm. In yet another paradox, both increasing the allocation of resources and constraining them have stimulated innovation. Ingenuity has sometimes been a child of hunger, leisure, or frustration, of the response to a stimulus, or a product of having been relieved of a specific preoccupation. Although some military inventiveness has flowed from focused research within defense industries, techniques and ideas have also been culled from the larger corpus of science and technology. When the funding of warlike research was vastly reduced in the early nineteenth century and again in the 1920s and '30s, the *corpus scientiarum* continued to grow as private and industrial laboratories and independent inventors generated a flow of fresh concepts and devices. From that proverbial stream, military professionals, like fishermen on the shore of a great river, drew out various ideas and devices and turned them to their own purposes. The emphasis on technical education in the American military academies and other officer training programs has usually been seen as a function of the need for their graduates to understand, maintain, operate, and buy technical systems, but it has also aided them in monitoring of science and technology in lean times. Nor did innovators wholly ignore the military potential of their work. Some, like the Wright brothers, hoped their efforts would have a benign influence and not be harnessed to darker purposes, but such visions were frequently confounded.

As with other applications of complexity theory, the divergent patterns that appear from considering its relationship to creativity resemble the seemingly infinite branchings of a fractal computer program. Most important from the standpoint of war dynamics and military history is the dynamic of ploy–counterploy, in which adversaries each try to impose on their foes the burden of having to cope with extra degrees of complexity through innovation and novel stratagems, while striving to reduce the intricacies of warfare to simpler forms. At the very least, the meshing of complexity and creativity theories helps to explain why, in war, attempts to impose patterns and to simplify have so often been confounded.

NOTES

1. The recent increase in interest in the U.S. Army is noted in John B. Alexander, Richard Groller & Janet Morris, *The Warrior Edge* (New York: William Morrow, 1990), esp. pp. 28, 41, 44, and 177–185; for a perspective on the U.S. intelligence community, see Sherman Kent, *Strategic Intelligence for American World Policy* (Hamden, Conn.: Archon Books, 1965).

2. E.g., see Joseph D. Douglass, Jr. & Amoretta Hoeber, eds., *Selected Readings from 'Military Thought'* (Washington, D.C.: U.S. Government Printing Office, c. 1980) [U.S. Air Force Studies in Communist Affairs, Vol. 5, Pt. 2], esp. M. Ionov, "On the Methods of Influencing an Opponent's Decision," pp. 164–172; and A. Berezkin, "On Controlling the Actions of an Opponent," pp. 183–186.

3. Quoted in Caleb Carr, "The American Rommel," *Military History Quarterly*, 4:4 (Summer 1992), p. 77.

4. Quoted in Vance Palmer, *National Portraits* (Melbourne: Angus & Robertson, 1940), p. 191.

5. Salient cases include the Union army's failure to adopt repeating weapons, available from the onset, until the last months of the Civil War, and, in World War I, the use of poison gas, tanks, and Q-ships initially in small increments, thus forfeiting their potential for surprise.

6. For a recent model of creative behavior encompassing those elements, see Howard Gardner, *Creating Minds: An Anatomy of Creativity Seen Through the Lives of Freud, Einstein, Picasso, Stravinsky, Eliot, Graham and Gandhi* (New York: Basic Books, 1993). Paradoxically, the military have tended to go counter to fashions of dress among *bourgeois* males since the French Revolution, and follow the patterns in nature and preliterate societies in which males display the gaudier plumage.

7. See Catherine R. Stimpson, "Lives of the Geniuses," *New York Times Book Review*, October 10, 1993, p. 37.

8. For a consideration of the implications of chaos theory for the academic study of international relations, see Yale H. Ferguson & Richard H. Mansbach, *The State, Conceptual Chaos, and the Future of International Relations* (Boulder: L. Rienne, 1989), esp. pp. 82–83 and 99.

9. See Clifford A. Pickover, *Computers, Patterns, Chaos and Beauty: Graphics from an Unseen World* (New York: St. Martins, 1990), p. 379.

10. Hence John Erickson's classic triad for analyzing military institutions of "technology-doctrine-style."

11. So cautioned the framers of the "garbage can model of decision-making," James G. March, Roger Weissinger & Baylon and Pauline Ryan, *Ambiguity and Command: Organization Perspectives on Military Decision-Making* (Marshfield, Mass.: Pitman Publishing, 1986), p. 18.

12. J. C. Wylie, "Why a Sailor Thinks Like a Sailor," *U.S. Naval Institute Proceedings*, 83:8 (August 1957), p. 813.

13. Robert McQuie, "Military History and Mathematical Analysis," *Military Review*, 40:5 (May 1970), p. 8.

14. *Condensed Analysis of the Ninth Air Force in the European Theater of Operations* (Washington, D.C.: Office of Air Force History, 1984) [orig. publ. 1946], p. 48.

15. R.V. Jones, "The Natural Philosophy of Flying Saucers," in Edward U. Condon, *Final Report of the Scientific Study of Unidentified Flying Objects Conducted by the University of Colorado Under Contract to the United States Air Force* (New York: Bantam Books, 1969), p. 925; for an analysis of misperception in another stressful context, see L. Hendrickson & J. Myers, "Some Sources and Potential Consequences of Errors in Medical Data Recording," *Methods of Information in Medicine*, 12:1 (January 1973), pp. 38–45.

16. Frank Barron, "The Needs for Order and for Disorder as Motives in Creative Activity," in Calvin Taylor & Frank Barron, eds., *Scientific Creativity: Its Recognition and Development* (New York: John Wiley, 1963), p. 160.

17. John Hogan, "Profile: Reluctant Revolutionary: Thomas Kuhn Unleashed 'Paradigm' on the World," *Scientific American*, 264:5 (May 1991), p. 40.

18. Edward de Bono, *Practical Thinking* (London: Jonathan Cape, 1971), p. 73.

19. P.R. Nayak & J.M. Kettringham, "The Fine Art of Managing Creativity," *New York Times*, November 2, 1986, p. F 2.

20. Alfred F. Dewar, "The Reorganization of the Naval Staff, 1917–1919," *Naval Review*, 9:9 (September 1921), p. 187.

21. David Bohm & F. David Peat, *Science, Order and Creativity* (Toronto: Bantam Books, 1987), p. 229.

22. *Impact of Air Attack in World War II: Selected Data for Civil Defense Planning: Division III: Social Organization, Behavior and Morale Under Stress of Bombing*, Vol. 1 (Stanford, Calif.: Stanford Research Institute, 1953), p. 115.

23. A well-known concept in that respect is Anthony F.C. Wallace's "Revitalization Movements," *American Anthropologist*, 58:2 (April 1956), pp. 264–281.

24. Constance Holden, "Creativity and the Troubled Mind," *Psychology Today*, 3:4 (April 1967), pp. 9–10.

25. Robert Sherrod, *Tarawa: The Story of a Battle* (New York: Pocket Books, 1944), p. 59.

26. For recent discussions of such concerns, see John Ellis, "Panic and Mutiny," in his *The Sharp End: The Fighting Man in World War II* (New York: Charles Scribner's Sons, 1980), pp. 255–270; and Richard Holmes, "Taking the Strain," in his *Acts of War: The Behavior of Men in Battle* (New York: Free Press, 1985), pp. 223–244.

27. For perspectives, see Henry Steele Commager, ed., *The Blue and the Gray* (Indianapolis: Bobbs-Merrill, 1950), pp. 976–982.

28. E.g., see Chester Wilmot, *The Struggle for Europe* (New York: Harper & Brothers, 1952), p. 588.

29. Jean Lacoutre, *De Gaulle: The Rebel, 1890–1914*, trans. Patrick O'Brien (New York: W.W. Norton, 1990), p. 72.

30. E.g., the U.S. Army decided not to published its projected volume on the Military Police Corps in World War II.

31. Some well-known examples are Nelson's putting his telescope to his blind eye at Copenhagen in 1805; Yorck's defection at Tauroggen in 1812; McClellan's open contempt for Lincoln, 1862–63; the Curragh Mutiny of 1914; Billy Mitchell's defiance of Coolidge, 1925–26; leaks of classified data by British officers to the press in World War I, and by a U.S. naval officer in 1949; Halsey's bolt to the north at Leyte Gulf in 1944; the political activism of Japanese junior

officers in the 1920s and '30s; the anti-Nazi activities of some German officers, 1936–44; and MacArthur's ignoring of Truman's directives in 1950–51.

32. Bohm & Peat, *Science, Order and Creativity,* p. 53.

33. Stephen Hawking, *A Brief History of Time: From the Big Bang to Black Holes* (Toronto: Bantam Books, 1988), pp. 9–10.

34. William Slim, *Defeat Into Victory* (London: Cassell, 1956), p. 209.

35. Alan C. Newell, "The Dynamics and Analysis of Patterns," in D. Stein, ed., *Lectures in the Science of Complexity: SFI Studies in the Science of Complexity* (New York: Addison-Wesley Longmans, 1989), p. 107.

36. Hence the myriad of popular pejoratives referring to displaying intelligence and generating ideas, e.g., "wise guy," "smart-ass," "know-it-all," and "egg head."

37. Boris Teplov, "K Voprosy O Praktischeskom Myshlenie," *Uchenye Zapiski,* 5 (1945), pp. 149–214.

38. A recent perspective is Russell Ruthe, "Trends in Non-Linear Dynamics," *Scientific American,* 268:1 (January 1993), p. 140.

39. See John Arquilla & David Ronfeldt, *Cyberwar Is Coming!* (Santa Monica: RAND Corporation, 1992) [P-8891].

8

Chaos Upon Chaos:

The Environmental Impact of War

While perspectives on warfare change when viewed through the conceptual lenses of chaos-complexity, that is especially true in respect to the influence of subtle forces, and more so in examining war's effects on the environment. Although the ecological impact of military operations was often visible in the flow of history, and sometimes dramatically so, relatively little attention was paid to it until very recently. There were, of course, some early peaks of concern, after World War I, toward the end of World War II when western Allied forces entering Germany were shocked by the rubble created by strategic bombing, and during the Vietnam War. Most recently, during the Gulf War of 1991, the dramatic televised images of the environmental effects of warfare were conveyed across much of the world, and literally tens of millions of people witnessed the Dantean hellscape of the Kuwait oil field fires.

While concern for "collateral damage" increased throughout the twentieth century, depiction of it has been uneven. In journalism and military history, war damage often appears as a background theme in illustrations of military subjects, or as statistics of military casualties, weapons destroyed or captured, and territory gained or lost. Occasionally, numbers relating to the destruction of railways and factories, and civilian losses due to bombing raids or blockades have been reported, but overall, "collateral damage" has been left off to the side, deemed incidental, or presented anecdotally or as sensational incidents. Deliberate slaughter of civilians as an act of war has been noted in military historical narrative when records or witnesses of it survived, as in the cases of Guernica, Katyn Forest, or the massacre at Oradour-sur-Glane, but many veterans will bear testament to much incidental damage and excess having been overlooked or ignored. The fragmented view of news media and the tendency to focus on immediate military consequences of warfare were visible in the Gulf War. The statistics of military casualties were seen as much "firmer" than civilian losses, and, aside from the scenes of the dramatic hit on a shelter in Baghdad, there was relatively little conveyed of the vastness of "collateral damage," and the fact that dozens of Iraqi civilians died for each Coalition military casualty, in spite of major attempts on the part of Allied airmen to limit such losses.

While that was obviously a product of censorship and limited access, it did little to alter the general view, especially in the United States, unmarked by war for almost a century and a half, that most of war's damage is done on battlefields. That is reflected in the recent flurry of concern about commercial developments near Civil War sites, and the frequent reference to them as "sacred ground." The scope of war damage is appreciated mainly by those who have suffered from it, or those who have undertaken an investigation of it. Detailed descriptions of the human suffering and physical damage caused by war can be found in fiction, memoirs, journalistic accounts, pacifist literature, the publications of international relief agencies, the Carnegie Foundation for Peace reports of the early twentieth century, such movies as *A Time to Love and a Time to Die, Hope and Glory, The Third Man,* and a number of the post–World War II Italian realist films. Just as most of those have focused on human loss of life and health, most international and military laws, codes, regulations, treaties, customs of war, and rules of engagement have focused on limiting specific effects, and most especially on the protection of human life, health, safety, and certain categories of property, and have been less concerned with the environment per se, although that, too, has been changing in the domain of chemical, biological, and nuclear-related limitations. Beyond those filters are the normative formats of statistics in history textbooks that portray wars' impact in terms of military casualties and combatant nations' financial expenditures.

While the question of intent gets lost in numerical summaries, that is also irrelevent to those who suffer the effects of military weaponry. Aside from clear-cut cases like the shelling of towns by warships and air bombing of cities, it is often hard to tell when "collateral damage" has been deliberate, and where the boundary lies between atrocity and accident. While a tremendous amount of environmental destruction has occurred as the result of warfare throughout history, and in that sense the Kuwait oil field holocaust was only one bead on a very long string, technological evolution has been amplifying human destructiveness increasingly since the mid-1700s. Of all the various waves of concern regarding the effects of war on the environment, the most enduring was the surge in ecological concern in the United States and western Europe during the Vietnam war, one of the most enduring attitudinal changes of that period.[1] Yet it is relatively recently that the broader historical pattern of war's damage to the environment has come under close scrutiny.[2]

During the Vietnam War, many saw "collateral damage" and deliberate attacks on the environment as relatively unique to that conflict and by-products of militarism, although many of those particular methods and initiatives were produced by civilian technicians and approved by civilian leaders. While some critics opposed specific ways of waging war, and others the Vietnam War specifically, or war per se, few were prepared to accept the military professionals' *defensa* that such weapons protected their forces at an acceptable environmental cost. Beyond citing tactical utility, some defenders of U.S. involvement and methods in Vietnam saw critiques of such methods as defoliation and the use of napalm as attempts to wrap antiwar arguments in scientific terminology and scholarly apparatus. On the other hand lay the wide array of technological "fixes," including the Rome plow, the King Ranch chain, napalm, and defoliation, along with the massive air-bombing and firepower made visible by media to vast audiences around the world, which many perceived as technological bullying and "overkill."[3]

Many American military and government officials, including those otherwise supportive of American involvement, criticized such techniques as the products of overdependence on "hardware" and symptomatic of a general lack of coherent policy. The tangled nexus of organizations, the bewildering arrays of equipment and ill-formed strategies and concepts were seen as complexity or chaos per se, or as the consequence of faulty organization and policies. Assessing the effect of any particular system or device of the many dozens employed was complicated by the dynamic of the Vietnam War, which saw very few major sustained battles or fixed lines of contact, but was mainly a fluidic and fragmented montage of flickering and formless patterns that thwarted the attempts of the Americans and South Vietnamese to plot trends and develop a coherent format. Immersed in that ambiguity, some military professionals hungered for the relative linearity of earlier conflicts and freedom from constraints on the use of force.

While the formats of ground combat in World War II were coherent relative to Vietnam, they had been less so than the tactical gestalts of earlier wars. Some analysts traced the trend toward assaults on the environment of an enemy in lieu of seeking of battle with enemy forces to Sherman's March to the Sea.[4] Whether or not the roots of the American military professionals' predisposition to use linear tactics and industrial methods in Vietnam ran that far back in history, they were highly visible in World War II and Korea. Those elements were reinforced in the early 1960s, as Secretary of Defense Robert McNamara and his entourage of systems analysts—the so-called whiz kids—imposed a strong ethic of numerical measurement on the American military nexus. That led to special complications in the Vietnam War, as the American military attempted to precisely measure operations and to force regular patterns on the episodic fluidity. The pressures running from Washington to produce results in a format derived from industrial production and sales statistics led many field commanders, most of them trained as engineers at West Point, to fixate on such indices as "body-counts," sorties flown, and ordnance "delivered" as primary measures of operational effect. The realization in fighting units that such "numbers games" were meaningless in limited war, but crucial in career dynamics, led to "cooked" figures, "fudge factors," and inflated estimates. That drove the assault on the environment, direct intent aside, as a side effect of delivering massive amounts of shells and bombs. Within a numerical ethos focusing attention on statistics of enemy dead and desertions, both environmental devastation and its impact in the domain of psychological warfare were treated as vague elements in a very "fuzzy" equation of warfare.

While there was some sensitivity in U.S. government circles throughout the war about such dimensions yielding negative propaganda effects in Vietnam, and on public opinion in the United States, and the world at large, there was less concern about environmental consequences per se. Many postmortems of the Vietnam War reflected frustration at being unable to mold the bewildering collage of events, motives, and so on, into a coherent configuration, and at not being able to wade in with the proverbial gloves off. Considering the environmental side of that war in the light of complexity and chaos offers a different perspective, aside from the fact that Vietnam, if plotted on a spectrum of warfare running from minimum complexity to chaos, would be seen as lying far closer to the latter than the former.

Instances of both deliberate and environmental devastation can be found in ancient accounts, including the ravagings of the Israelites and their foes, especially those of the Assyrians, well known to students of the Bible, while Classicists are familiar with the Romans' sowing of salt on the ruins of vanquished Carthage. Of the many barbarian hordes who ravaged Europe, the best remembered are the Huns, whose chief, Attila, purportedly proclaimed that nothing would grow where his horse had trod. Substantial indirect effects of war on ecological balance, less vivid but perhaps greater in effect, included Athens' denuding of Attica's timber for warship construction, and the Mongols' shattering of the Khwarizmian Empire's delicately balanced canal networks, which returned much of Iran and the Tigris-Euphrates Valley to the desert. In the Middle Ages in western Europe, deliberate environmental depredation by feudal levies took the form of the *chevauchée*, great foraging sweeps aimed at gathering subsistence while devastating enemy lands. That line of logic was reversed in the "scorched earth" tactics of Bertrand du Guesclin in the final phase of the Hundred Years' War, by the Russians in 1812, including the burning of Moscow, and again in 1941 as they fell back from the Wehrmacht's onslaught, and in the Sino-Japanese War, 1937–45.

The spirit of the *chevauchée* was apparent in various American wars, including devastating raids and punitive expeditions along the northern frontier by both sides during the American Revolution. It flowered full-blown in the Civil War, in Sherman's "March to the Sea," as a 70,000 man army cut a sixty-mile-wide swath of destruction across Georgia from Atlanta to Savannah in late 1864. Much in the same spirit was Sheridan's savaging of the Shenandoah Valley that destroyed the base area of partisans such as Mosby's Rangers, denying the region's agricultural riches to Confederate forces, and ending the threat of a sudden descent on Washington. Although the slaughter of the great buffalo herds on the Great Plains a decade later was not a military effort per se, it had military effects. General Sheridan sanctioned it, and later judged it a root cause of bringing down the Plains Indian tribes.

The vast swaths of destruction cut by the fighting armies in World War I were less deliberate, but far greater in scale. As huge armies deployed on five major fronts and in several theaters of war, supported by vast webs of railways, steamship lines, and mass production, the earth was widely scarred by battle, production, construction, and extraction. All the major combatants' war planners had anticipated a conflict of a few months at most, and as their industries were mobilized in stages over the next four years, millions of shells fired by thousands of guns and hundreds of gas attacks produced unforeseen "devastated areas," some portions of which have remained unreclaimed since 1918.[5]

In both World Wars, the bite of warfare on forests and jungles deepened, both in the form of battle damage, and in the search for resources, some of it a "follow-on" to the aggressive harvesting of woodlands for naval construction materials from the Classical age to the mid-nineteenth century. For example, during World War I, vast giant spruce stands in the Pacific Northwest were wiped out for aircraft construction. A number of schemes for igniting enemies' forests were also proposed by military leaders and planners and by civilians. In 1915, when a citizen of Birmingham sent a scheme to Field Marshal Sir Douglas Haig, commander of the British Expeditionary Force, to drop four million fire pellets on Germany in a single night, the proponent pointed out a special advantage: no alarm would be raised since their landing and detonation made little noise. At the same time, a Canadian's proposal for a type of gasoline bomb for attacking forests led to lengthy (and sometimes silly) deliberations in the British defense bureaucracy without result.[6] A plan drawn up in mid-1918 for using fire balloons also generated some interest, but to no result.[7] Aerial attacks on agricultural targets were also envisioned at that relatively early point in the "air age," including a biological "vector," in the form of a scheme for dropping Colorado potato beetles on German potato-growing regions.[8] Later in the war, General William "Billy" Mitchell, the American air power enthusiast, joined in the spirit of things in offering proposals to set German crops and forests afire and to use poison gas against livestock.[9]

The deepest scars inflicted on the environment were the consequences of waging war as a mass industrial enterprise. Although most toxic residues, deforestation, and the blasting of millions of craters along the major fighting fronts in World War I resulted from military operations aimed at other purposes, a major exception was Operation ALBERICH, carried out in March 1917 by the Germans as they withdrew eastward from the Somme battlefields into the Hindenburg Line. Named for the nasty gnome in Wagner's *Das Rheingold,* ALBERICH led to the systematic savaging of a zone of sixty-five by twenty miles in which buildings were flattened, wells poisoned or blocked, bridges destroyed, and trees cut down.[10] Soon afterward, the extended mass artillery bombardments preceding British attempts to break through in the Battle of Passchendaele yielded a similar effect, although that was not the direct goal. The plan for millions of shells to be fired by thousands of heavy guns was opposed by British army engineers, who warned British Expeditionary Force headquarters that very regular seasonal rains would inundate Flanders if its delicately balanced canal system was shattered. Their predictions proved correct, and the Ypres sector became a vast complex of lakes, swamps, and thick mud.[11]

In World War II, the environmental effects of warfare, more widespread and greater in scale than in World War I, were also the product of a mix of intentions and inadvertent effects. In Britain, visions of setting forests alight[12] and attacking crops with insects reappeared,[13] but again, were not carried through to praxis. Neverthless, much damage was done to the forests of central and eastern Europe, both by Allied bombing errors and by the Nazis' diverting the Royal Air Force's night bombers by igniting fake target-marker fireworks in the countryside. Aerial attack schemes aimed at causing major environmental damage included the much-publicized RAF "Dam Busters" attack on the Sorpe, Eder, and Mohne dams in the Ruhr in 1943, which resulted in widespread flooding. Less well known were the Japanese windborne incendiary balloon bomb attacks on American and Canadian western forests, which had minimal effects.

A more diffuse but substantial source of pollution outside industrial sites and mines was the spills of oil caused by the sinking of thousands of merchant vessels and warships in the world's oceans and the Mediterranean Sea, as well as hundreds of oil, gasoline, and chemical tankers. From 1940 to 1945, about a hundred major cities in Europe and Japan literally went up in flames from incendiary raids in conflagrations that included many thousands of fuel storage tanks and dozens of chemical refineries and plants. In China, Russia, the Balkans, and western Europe, many thousands of villages, towns, and cities were extensively burned in the course of "conventional" military operations. At the same time, the maneuvers of vast mechanized armies, air forces, and navies added to the emanations while tearing up the soil. Especially dramatic was the churning up of vast dust clouds by thousands of motor vehicles and the breaking of the desert's crust in the North African campaign, traces of which remain today. In another dimension, DDT insecticide was used in vast amounts as a topical disease preventative, and in some aerial spraying, including "applications" to very large areas before assault landings in the Pacific theaters of war. The greatest individual acts of deliberate environmental destruction, however, were the deliberate floodings of vast areas in China by the Nationalists in 1938, in the Ukraine by the Soviets in 1941, and of Holland by the Germans in 1944.[14]

The environmental effects of military operations since World War II can only be estimated. The dozens of "small wars" that have raged at any one time during the cold war have continued as a kind of sinister, grisly background static in the "New World Order." In some of those conflicts, environmental damage was plainly visible, as in the slashing of rubber trees in Malaya in the 1940s and '50s. Some long-ignored consequences of past struggles have subsequently emerged, like the massive implanting of land mines and booby traps throughout the world that continue to claim victims.[15] Those and the ongoing revelations of past and ongoing militarily-related pollution and the proposed disposal of weapons make future prospects unclear.[16] Soaring estimates of the costs of closing military bases and bringing them to environmental standards have brought the dilemma of delayed effect into very sharp view.

Technical evolution in the military sphere may offer some remedy, but that is not at all clear. The development of highly accurate missiles, shells, and bombs may reduce incidental devastation, but those ameliorations are being offset by development of such devices as biochemical and genetic weapons, and fuel-air explosives. Although prospects of a "nuclear winter" seem far less than they were when Jonathan Schell limned his lyrical and ghastly scenario in *The Fate of the Earth* in the early 1980s, the fading of superpower rivalry has not halted horizontal nuclear, chemical, or biological weapons proliferation. Nor is it clear that the new order of things has reduced the likelihood of their being used in war, since the fear of escalation and catalysis in the cold war that served as a control is gone.

Beyond identifying specific problems requiring immediate remedy, what value is there in more closely measuring war's impact on the environment? A truly exhaustive tracing of that would be a daunting exercise. Not only would many interlocking orders of complexity that defy easy identification be encountered, but an obscuring effect that arises from the commonsense view that many such "wounds" have been healed and smoothed over the years, decades, and centuries by the recuperative powers of humankind and nature, an argument that looks something like a "covering law." Viewing the impact of war on the environment using complexity theory offers little comfort, since it raises a possibility that further surprises might emerge as the great outcomes of unseen or very small and seemingly inconsequential "initial conditions" or subtle influences. On the other hand, all the environmental damage done by weapons, the maneuvering of forces, the spillage of fuel and lubricants, and cratering may ultimately prove to have been a side issue in history, military and otherwise.

It is not clear if an appraisal of ramifications would exclude or overlook critical subtleties, or if the identification of those is literally beyond human capacity. There has already been a good deal of shying away from what was seen as too complex, irrelevant, or unpleasant, but which was nevertheless pregnant with significant environmental consequences. In the darkling world of nuclear war theory and planning, analysts and designers focused on damage done by "nukes" to "strategic" (i.e., nuclear-weapons–related) targets, whereas other effects, such as the random impact of diverted or faulty weapons or the spillage of nuclear material as "nukes" hit "nukes" were seen as extraneous and left outside the borders of consideration by using "cookie cutter" models and equations, even though the aftereffects of Hiroshima, Nagasaki, and hundreds of atmospheric nuclear-thermonuclear weapons tests emerged into public view from time to time. Given the recent revelations of bizarre experiments of the 1950s and '60s by U.S. government agencies, it would be very rash to suggest that those experiments are all behind us.

Placing the continued emergence of such seemingly marginal side effects in the context of complexity theory makes it discomfiting to consider that such darkling scenarios of environmental meddling as *The Island of Doctor Moreau* and *The Andromeda Strain* were painted on large figurative canvases, depicting vivid deviations from the norm, or in the case of *On the Beach* and *The World, the Flesh, and the Devil,* vast cataclysms. Those novels and films were based on the model of scientific meddling and technology running amok and generating obvious, large-scale cataclysms, not the insidious accretion of subtle shifts in the *Ganzfeld*, like Rachel Carson's *The Silent Spring*.

Both the spectacular environmental devastation in the Gulf War and the outbreak of mysterious ailments among veterans brought back into view the question of environmental side effects caused by evolving nonnuclear military technology by design, accident, or both. They also touched on the quandary of how much attempts to protect the environment against pollution and despoliation may be offset or negated by effects of war or the preparation for it. Considering those dilemmas with complexity in view casts a different light on cumulative effects. Although indistinct and difficult to measure, such extraneous events may have had far more impact than is generally appreciated, even though, as with other kinds of environmental scarring, concern has been dulled by the mythos that the vastness and complexity of nature itself blunts such effects and aids recuperation. At a further order of complexity, it is not clear how much warfare's environmental impact has intertwined with other vectors such as slash-and-burn agriculture, toxic industrial processes, population growth, and resource depletion to form a grim synergy. If theories of complexity regarding sensitivity to initial conditions are applicable, not only major, visible effects of warfare, but minor, obscure influences might yield disproportionately large state-changes, or may have already done so in as yet imperceptible ways, either as catalysts, as sub-elements of a wide spectrum of forces, or as kinds of proverbial straws in the cumulative burdening of the ecological camel's back. Given the stakes and the uncertainties, it is not surprising that many on both sides in the environmental debates of the last generation have taken a position of moral exclusivity, which is usually the case when humans argue untestable hypotheses.

What, then, are the uncertainties? Several obvious questions arise from the aligning of chaos-complexity theory with the environmental impact of war. Has much been done that is not yet apparent, but that may become visible later, or never seen as such, but which will have profound effects nonetheless? Was damage done in specific cases or overall damage that was visible less modified by natural process and time than was assumed? Have aggregates of such destabilizing forces been forming, the slight stimulation of which might trigger a massive, sudden state-change? As such constructs lead to a more careful tracing of the roots of war and human assaults on the environment, and of their ramifications, which may prove to be far more serious than previously imagined.

NOTES

1. E.g., see Arthur H. Westing and E.W. Pfeiffer, "The Cratering of Indochina," *Scientific American*, 226:5 (May 1972), pp. 21–29; and E.W. Pfeiffer & Arthur H. Westing, "Land War," *Environment*, 13:9 (November 1971), pp. 2–13.

2. E.g., see Arthur H. Westing, ed., *Environmental Hazards of War* (London: Sage Publications, 1990) and Susan D. Lanier-Graham, *The Ecology of War* (New York: Walker & Co., 1993).

3. For details of such methods, see *Field Manual 5–164, Tactical Land Clearing* (Washington, D.C.: Department of the Army, 1974).

4. E.g., see Russell Weigley, *The American Way of War* (New York: Macmillan, 1973), p. 152.

5. For details on the ongoing efforts to counter those effects, see Donovan Webster, "The Soldiers Moved On. The War Moved On. The Bombs Stayed," *Smithsonian*, 24:11 (February 1994), pp. 26–37.

6. PRO (British Public Record Office, Kew) AIR 1/21–15/1/100, "Destruction of Forests by Bombing, Proposals for," 17.7.17–1.3.18.

7. PRO AIR 1/1985, "Report Correspondence re Questions of Using Balloons for Carrying Incendiary Bombs for Destruction of Enemy Crops," May–August 1918.

9. Isaac Don Levine, *Mitchell: Pioneer of Air Power* (New York: Duell, Sloan & Pearce, 1958), p. 148.

10. See B.H. Liddell Hart, *The Real War 1914–1918* (Boston: Little Brown, 1930), pp. 299–300; for a contemporary, propagandistic account, see *Frightfulness in Retreat* (London: Hodder & Stoughton, 1917).

11. Leon Wolff, *In Flanders Fields: The 1917 Campaign* (New York: Viking Press, 1958), pp. 81–82 ff.

12. E.g., Reinhard Gehlen, *The Service: The Memoirs of General Reinhard Gehlen* (New York: World Publishing, 1972), pp. 104–105.

13. PRO AIR 14/1929—Memo, Portal to Peirse, 15 May 1940.

14. Westing, *Environmental Hazards*, pp. 6 and 11.

15. Donovan Webster, "One Leg, One Life at a Time," *New York Times Magazine*, January 23, 1994, pp. 26–33, 42, 51, 52, and 58.

16. E.g., see program brochure of the 20th Environmental Symposium and Exhibition, "Department of Defense Environmental Security Strategies for the 21st Century," March 14–17, 1994 (Arlington, Va.: American Defense Preparedness Association, 1994).

9

Considerations and Conclusions

Concluding a study based on complexity is paradoxical in itself. The sequential unfolding of dimensions, and especially dilemmas and paradoxes, works against focusing analysis and thwarts the drawing of neat conclusions. Whether or not looking at warfare through nonlinear lenses reveals actual fractal patterns, the unfolding divergences make history appear less like telling time by looking at a watch than looking at its inner works. Not only would pursuing such minutiae be dauntingly tedious for historians and analysts, but such exhaustive chronicling would have a very small audience, although some military historians have grappled with the minute mechanical details of warfare, usually with an antiquarian or technical bent. Increased awareness of subtle influences might lead historians to pursue such particulars more ardently, but the ethos of solitary research and authorship is deeply ingrained in the craft of history, even if the study of war, like the sciences, engineering, and social science research, has moved beyond the capacity of individuals to do justice to its intricacies. Whatever the case, it is hard to say where the tracing of such branchings might lead.

Another dilemma that nonlinearity raises is whether it really offers historians much more than a trove of metaphors and sensitivity to complexity generally, and to stochastic patterns and probability as elements of war and battle, awareness of which is not wholly novel in military historiography. Might it at least help explain why efforts to develop principles and laws have so often fallen short of being predictive aids, or failed to lead to the framing of valid nostrums?

The following discursions were designed to show how sensitivity to complexity opens ranges of consideration. Running against strong tides of convergence and patterning in history, it generates splays of quandaries, dilemmas, paradoxes, and ambiguities. These ruminations may stimulate further consideration of the problems identified, or of other issues in history worthy of more detailed examination, or a reexamination of assertions made on the basis of linearity, selectivity, and simplification. If they merely irritate, hopefully it will be in the spirit of Charles Saunders Pierce's "irritation of doubt."

INTELLIGENCE PROCESSES: COMPLEXITY MASKED AND FILTERED

Over the last half-millennium in the West, and far longer in such places as China, the sensing of the complexity and chaos of war has been the product of deliberately structured intelligence networks that enfold a paradox. While they are designed to reduce ambiguity and confusion and provide leaders at the focus of the flow with clarified views of uncertainty, the shields and compartments in those systems designed to protect those with supreme authority from monitoring and ostensible linkage of spies and saboteurs also mask and filter that information. That shroud of secrecy has lent mystery and glamour to public images of espionage, even though much intelligence work is pedestrian, bureaucratic routine such as cutting up newspapers and scrutinizing telephone books. Nevertheless, most popular portrayals of the "business" have focused on historical cases, like Mata Hari and "The Man Who Never Was," or have been lurid fictions like the James Bond and Matt Helm series. As with detective stories, those have emphasized puzzle-solving, spying, monitoring, sabotage, and deception. Since World War II, historians and policy scientists have increasingly scrutinized synthesis, interpretation, and shaping estimates, whereas fiction writers, most notably Len Deighton, Graham Greene and John le Carré, have probed uncertainty of effect, ambiguous allegiances, cynicism, and bureaucratic dynamics.

The intelligence process is generally viewed as an upward, converging flow of fragments of information, gathered by agents within cells separated for secrecy (HUMINT), or by remote sensing techniques (TECHINT/ELINT/SIGINT). Synthesis, matching, and deduction are involved at all levels, but complexity increases toward the top of that flow, especially since spurious data ("noise") increases there as analysts practice a form of creative artistry, breathing dimensionality and life into the fragments—but within a highly strictured organization, designed to mold or "digest" fragments into coherent information, and channel, expedite, distribute, and protect it. As with other "dialogue(s) between harmony and invention," crosscurrents between individuals and structure will, ideally, be balanced, hence the interest in creativity theory that arose in the American intelligence community in the late 1940s. Since then, advances in sensor and computer technology, satellites, photogrammetry, cryptography and cryptanalysis have altered the intelligence "business," but not the basic cycle of gathering information, synthesis, analysis and estimation, and conclusion that follows the sequence of creativity. Some aspects of that process are generic among military and civilian intelligence agencies, but the substantial differences from case to case are in keeping with the Prussians' "law of the situation." "Combat intelligence," "command information," and "strategic intelligence" may differ widely in one instance and substantially overlap in another.

Aside from those technical advances, evolutions in military and naval command-and-control have also affected intelligence processes. The military and intelligence communities overlap heavily and were mutually affected by increases in encryption and decryption, and volume and rate of data flow in networks, causing overload, rising expense, and complications in organization, analysis, routing, and security. The persistence of some qualitative problems in both domains became visible when, as the air offensive in the Gulf War began in January 1991, difficulties in analysis and circulation surfaced.[1] Such intelligence "glitches" had appeared throughout the cold war, and others have surfaced as superpower tensions have ebbed.[2] Considering those with complexity and chaos in view leads to further quandaries. For example, can observers—or overseers—outside the inner workings of the "business" know that such gaffes are real, or if they masked or diverted attention from successful ploys? It has often been claimed that intelligence successes have outweighed failures, and the latter are normative "breaks of the game." But that leaves it unclear how certain "insiders'" perceptions are. Can success or failure really be measured as precisely as claimed?[3] Can historians ever judge that? Many have maintained a strong faith that valid "paper trails" exist that will ultimately allow evaluation (i.e., the "judgment of history") despite the many visible cases of "sensitive" data being eroded by technology, destroyed, or suppressed.[4]

Beyond those distorting lenses, the historical rendering of intelligence activities is affected by other blurring forces, including the reflexive imposition of patterns and logics, and the impulse to reduce or discard ambiguities, or bolster them with deduction and inference. Direct access to very sensitive and therefore crucial material is generally constrained. Official histories are screened to some degree, and although direct editorial pressure may be minimal, authors are chosen and source access may be modulated. Historians working outside government and military structures, however they strive for "objectivity," may be affected by bias, the desire to please a particular interest, or market trends. Their access to evidence may be affected by their reputation, contacts, or inclination of functionaries to help them on their way. Nevertheless, the historical profession and many "consumers" assume that reasonable if not absolute truth exists in the form of accessible evidence, from which a facsimile can be assembled by competent and insightful researchers rigorously applying appropriate methodology. Yet there are further complications. If, for example, substantial new evidence became available about, say, the Bay of Pigs or the Iran-Contra affair, researchers would know what to focus on, having a perspective that decision makers and "actors" lacked at the time. Assuming such data was more than fragmentary, and held in archives, categorizing would very likely have extracted it from the turbulent flood of unorganized information (including telephone and personal conversations) that engulfed those who were engaged with that set of problems that were not then labeled as a historical "case," "affair," or "incident," or even seen as being linked. Hence there is a sense of those involved in the policy process of coping with a panoply of vague and dissonant elements closer to chaos than complexity.

Analogically, then, historians are intelligence analysts of the past. They, too, work with uncertain evidence, follow threads and clues, and are sometimes hampered by obfuscation or denial—but in calm, well-ordered research environments, without personal involvement or hindsight. That contrasts sharply with the intelligence domain, especially in wartime and crisis, where such cross-forces as stress, fatigue, fear, distortion, destruction, and deception violently warp research and analysis. Thus the intelligence process resembles meteorology in relying on scientific procedures and equipment, and in analysis, estimation, and prediction being less than a science. In both fields, despite advances in methods and computer power, there are "no recipes or formulae, no check lists or advice that describe reality . . .,"[5] although training, structure, and process are designed to minimize distorting influences. Those have not eliminated surprise blizzards, floods, and hurricanes, or, in 1989, the shock of the Iraqi invasion of Kuwait, or the sudden collapse of the East Bloc. Postmortems of such events, like Roberta Wohlstetter's *Pearl Harbor: Warning and Decision,* have shown how, in intelligence nets as in any complex organization, no one at any level can perceive more than a fraction of the entire organization's state at any moment, nor can that ever be fully sensed or depicted, even in hindsight. The angle of view and general sense of detail of the aggregate tend to be greater toward the top, but what happens throughout the organization is sensed only approximately. Details may seem much clearer at "the cutting edge," but only across a limited range, allowing little or no sense of general context. In intelligence organizations, that effect is heightened by compartmentation and division of crucial information to maintain secrecy.

Here, considering complexity theory as regards the disproportionate influence of small forces in complicated systems leads to yet another paradox: it is rarely clear if—or how much—one vantage point is better than another at the time of an event or in hindsight. That stands in sharp tension with common sense, and with many theories and cultural attitudes about perspective being a function of hierarchical altitude; directly and inversely, this quandary is reflected in arguments over the virtues of centralization and decentralization. At a further order of complexity, forecasting in all kinds of organizations takes place in a constantly changing milieu, influenced by distorting forces of various magnitudes, some invisible, others seemingly subtle or minute. Those include individual idiosyncracies, accidents, and turns of fate, in what Enoch Powell called one of the largest "'black holes' in history . . . the inner story of the emotions, including sexual dependences of its principal actors."[6] Such anomalies add a splay of fractals in charting events, since each viewer sees events through filters of prior knowledge, experience, state of mind, attitude, and temperament, from a unique vantage point, in keeping with the physicists' "law of location." Linked to that are other orders of "fuzzy sets," including the interplay of personalities, cultural and factional tensions, interpretation of bureaucratic rules, ideologies and biases, wishful thinking, technological change and resistance to it, individual and group reflexes, habits and mind-sets, and known or sensed desires of "consumers."

The cellular structure and extreme secrecy of intelligence networks cut across those intricate fans, limiting discourse and feedback and distorting fields of view. Such similes as intelligence process being like the assembling of a jigsaw puzzle fall short of conveying the sense of a *Feld* that cannot be *ganz*. If any pieces are in the box at all, they are unlikely to be those of a single puzzle. There may be no label on the cover, or it may be a false one. The box may contain pieces of many other puzzles, some intentionally misshapen or wrongly marked by an adversary. Assemblers will rarely be sure what picture they are trying to attain, but may have to guess that at any point during assembly. In some instances, important pieces may be known to be somewhere else, and have to purchased, or picked up amid the flow of traffic on a busy highway, or in the lion's den at the zoo, or may known only to be lying about randomly—and so on.

If, then, intelligence analysis is artful deduction based on indeterminable uncertainty, commanders and leaders also act on imperfect or sparse data they cannot verify, hence Burt Kosko's view of reality as gray, and theory as black and white.[7] Despite such blurring, the intelligence process has often been presented as more logical and routinized than it can be, or, perhaps, when considered from the perspective of creativity theory, should be. That is partly due to the extent to which it is seen as a major conduit of political as well as military power. Not only does the existence of an intelligence structure in itself burden adversaries with uncertainty about its capacity, but the arcane jargon and special atmosphere of furtive prestige arising from access to secrets adds to the mystique of leaders' and insiders' superior knowledge and infallibility. On the other hand, some historians and social scientists have treated intelligence as a faint leitmotif, a murky given, or a black box on an organizational chart or linear model. In spite of the apparently wide variations in the quality of intelligence that have been made visible, portrayals of major events and models of international relations are often based on assumptions of "rational actors" making careful and well-calculated decisions, engaging in bargaining, with goals, "payoffs," and "tradeoffs" well defined, and with key players possessing nearly complete information.

As with the dynamics of combat, applying complexity theory here reveals cracking and crazing at each level. Causally significant minor variations and discrepancies have often been identified by historians and journalists, who, like curious detectives, probed beyond accepted explanations to reexamine basic evidence and discover minor details, ignored or suppressed, that led to major reinterpretation. If complexity theory increases sensitivity to such factors, it might raise history in the view of practitioners, who often deem journalists' and historians' versions of events they experienced as simplistic, incomplete, or fanciful. Even when bureaucratic secrecy is not involved, historians can never be sure what goes unnoted because evidence has been destroyed or distorted to conform to ideology, assumption, or preference. In this murky area, complexity theory reinforces the skepticism that scholars, journalists, analysts—or leaders—bring to the perusal of documents, and to attributing causality, and the especially thorny quandary that John Lewis Gaddis called "the methodological impossibility of 'proving' inevitability in history."[8]

As difficult as that is with history in general, it is far more the case in diplomatic and military history, regardless of the strong faith of both "producers" and "consumers" that portrayal reasonably matches events. Despite the difficulties—or impossibilities—of verifying, a curious blending of history, journalism, and fiction over the last generation has produced a widespread sense of the intelligence process being visible and reasonably well understood. From the 1950s onward, many aspects of the intelligence "business" became public, like royalty during the same period. The mystique was somewhat eroded by letting "daylight in upon magic," both in revelations of intelligence failures and surprises, and by fictional portrayals of intelligence as futile and convoluted by Somerset Maugham, Evelyn Waugh, Graham Greene, and John Le Carré. Attitudes were shaped and arguments based on purported exposés and fictional imagery, although there was no way to weigh authenticity. The Church Committee hearings in the U.S. Congress and the purge of CIA covert action and espionage elements during the mid-1970s was partly due to the peaking of infatuation with "TECHINT" systems at that time. Yet much of the mystique of intelligence remained intact, because of secrecy and cruciality. Spying, a relatively small and often dull portion of the process, and wildly fanciful portrayals of "special ops" continued to receive far more attention in popular and fictional formats than analysis or estimation. Excitement faded with recognition that TECHINT systems could be spoofed or countered, and that "HUMINT" and "special ops" offered options that other modalities could not.

In the wake of the cold war, some judged intelligence to have been a marginal or sinister influence in that protracted struggle. While lack of access to data clouds the issue, complexity theory rolls that line of argument out into another fan of quandaries. Since all the factors bearing on the unfolding of events can never be known, the logic that half a loaf is better than none may be in error. Partial information, both in praxis and in analysis of events, might be worse than none at all. An implicit strategy shaped with that understanding in view would abandon attempts to gain the impossible or the unknowable, and rely on muddling through, a process that cannot be tested against alternatives. That gives little comfort at a time when international balance and "security" seemed to be declining at one level and rising at another. Paradoxically, the potential for Western powers' involvement in espionage or military operations has not declined as a direct function of the depolarization of the international *Ganzfeld,* Since that has yielded greater complexity, it can be argued that there is a consequent need for more perspectives, higher resolving power, and a greater sensitivity among analysts and "consumers" for shadings, catalysis, and so forth.

Which path is most likely to be taken? Most probably, neither. In spite of attempts in U.S. intelligence circles in the 1950s and '60s to mollify cramping effects of structure, routine, and organizational interplay, some aspects of process, such as bureaucratic rivalry, generate dissonance. Since that generally increases when resources are reduced, it does not seem likely that implications of complexity theory will weigh much in the balance against momentum, given the powerful resistance, at the "folk" level, to ascribing causation to blind fate or deeming events as inexplicable. That hunger for order and for rational explanation of anomalous or incomprehensible events is reflected in the proliferation of conspiracy theories—better a bizarre causal logic than none at all, or conceding that meaning is out of reach. Such deference to happenstance, like the Romans making decisions on the basis of auguries, is not likely to be accepted in the political arena, where "products" of intelligence are converted into policy and action. There, in American culture and some others, the myth of leaders having special personal powers is well enshrined, along with positive attitudes toward shedding details in defining and solving problems,[9] being a "quick study," focusing on the "bottom line," and seeking simple concepts and general norms.

If complexity theory became an active force in shaping organizational design, values, and attitudes, failure and error would be recognized as normative in doing "business," as would be the risk of depending on a small range of sources and analyses. There would be a greater appreciation of subtle influences in well-structured systems, such as informal process, careerism, small "p" politics, and the like, which have already been recognized by some journalists and historians. And there would be less concern for hierarchy, although a trend in that direction is already visible across the culture, in the vast on-line computer nets that allow rapid and fluid assemblage of ad hoc task forces or clusters. Participants in those firmaments address ongoing issues and special problems, and have come to constitute an inadvertent intelligence nexus beyond the wildest dreams of Sir Francis Walsingham or Allen Dulles—or the nightmares of Fouché, Himmler, Dzerzhinski, or James Jesus Angleton. It is reasonable to assume that the intelligence community cadges free rides on those networks, and has its own sophisticated counterparts.

Such prognostications really run counter to the implications of complexity that run against forming firm conclusions and toward a greater appreciation that what can be seen is rarely what there is. That is a weak guide in determining what may be, since what cannot be seen could be of far greater import than what can be. Complexity theory places secrecy in a different light, to the extent that security constraints suppress potentially significant influences. Thus disclosures of the CIA-sponsored drug and disease experiments, like the Soviets' bizarre nuclear exercises, may not have been anomalies or perturbations, but indicators of other unseen events, and what might ultimately happen on an equivalent or much greater scale.

"INVISIBLE HANDS" AND "BIG PICTURES"

Like Newtonian mechanics, Darwinism, and sociobiology, chaos-complexity theory might be used to support or refute arguments in the social and political realms, such as free-market dynamics versus centralized planning, or "ground truth" versus the "big picture" in command-and-control design. For example, the meshing of military and economic systems generates a wide range of intricacies and paradoxes. Long wars have increased restrictions, like rationing and blockades, while stimulating other types of development, innovation, and growth, including such exotic economic subspecies as black markets and smuggling. Restraints on government-vendor relations have sometimes been relaxed or abandoned, and government auditing and oversight, including low-bid procedures, suspended inside the boundaries of some clandestine activities. Contradictory patterns appeared during the cold war, confounding policy formulation and analysis. The United States had no formal national trade policy, but at various points, vital raw materials were stockpiled, patents deemed vital to national security impounded, lines of critical research shrouded in secrecy and/or subsidized by federal funding, and embargoes imposed. Beyond that, sizable shadow economies were created, enshrouded in "black budgets." As with the "big picture"–"ground truth" dichotomy, policymakers' and planners' awareness of that tier of complexities offers them no clear advantage in identifying and choosing options, or plotting patterns and trends. Nor does it help those who later strive to ascribe exact causes, make historical generalizations, or frame laws or principles.

In the realm of battle dynamics, disintegration of relative symmetry and linearity of combat tactics and formations over the last two centuries transformed combat into a collision of fragments, not wholly random but far less coherent than in the older formats. In complexity theory terms, most battles became multiple whorls around many attractors, not wholly random or chaotic, but thwarting the drawing of clear vectors and patterns. Have those trends been symptomatic of an interval of deficient organizational control that will ultimately be ended by technical improvements, or do they reflect an immutable property of international rivalry and warfare? Does complexity theory merely affirm that commanders, staffs, and combatants, as well as politicans, civil servants and diplomats, are doomed to be dismayed as actual operations splay out of alignment with their doctrines and plans? If that is true, is it a corollary that historians and journalists must continue to grope through uneven evidence as they strive to impose order, and be limited to figmentive description? What options are there for military professionals or other practitioners of force beyond mastering basic skills and trying to develop some tolerance for surprise, and a honing of their intuition and ability to judge quickly?

Those uncertainties, plus the likelihood of lean defense funding, point to a defense policy based on maintaining armed forces with a broad if thin array of skills, some of whose elements can react readily while others expand quickly and assemble contingency-specific "packages." In such a system, command-staff elements would consider many contingencies while forces trained, or at least planned to operate across a wide range of free-maneuver and surprise scenarios in many different regions.[10] Logistical support systems would aim at flexibility and adaptability and link with industry to be able to redesign, modify, and surge quickly from a low production base, or revive wholly dormant programs, as well as relying on civilian-military "dual use" procurement, and so on. A reverse of that model of expansibility would focus forces, resources, and expectations on a few assumptions and scenarios, gaining maximum familiarity and virtuosity within certain narrow ranges of what seem to be highly probable or very critical contingencies. Links with specific industries would be forged to assure greater mass and depth in those designated vectors, and close melding of people and equipment.

The former model roughly resembles the general defense policy of the United States between the World Wars, and the latter, its cold war policy. Clearly, it can be argued that something would be gained or lost by selecting one over the other, but without actuality in the equation, such debate is mere speculation. The most perplexing ideational branching here is the question of how much the choice of stratagem and policy actually shapes the eventual morphology of reality, that is, is it really true that preparing for war tends to suppress the probability of being involved in one? Fine-tuning expectations within the parameters of a focused doctrine could create false assumptions and expectations by excluding options, suppressing creativity, or implanting inappropriate mind-sets and reflexes. Conversely, a model based on wide and thin scanning could lead to superficiality, rashness, an unrealistic sense of realism about major operations, overreliance on reductionist simulations, and faith in simplistic visions. Either model also sends "signals" about intention and attitude. The examinations of doctrinal fragility showed how history can be "mined" to support various positions. Here, beyond underscoring implicit ambiguities, complexity theory grants no clear or easy solution. Some historical cases would be deemed assertions, and some would be judged as being irrelevant. Since judgments are subjective, and perspectives tend to grow "fuzzier" as more cases are brought into view and subjected to close, detailed scrutiny, planners looking for clear indices become frustrated.

The pragmatic, high-pressure environment of "real-world" bureaucratic defense planning and operations pushes process toward quick closure and apparently clear definitions, leading to simplifications and reliance on "cookie-cutter" logic such as "cutting to the chase." Since complexity theory tends to move the vantage point toward the seemingly extraneous, it seems unlikely that it will be given much credence, or that seemingly minor factors and contingencies will be more intensely scrutinized than in the past generally. Awareness of complexity-chaos does offer practically-minded leaders, planners, and analysts a clearer sense that they will, if they go to war, encounter uncertainties beyond anyone's ability to anticipate. While that might reinforce the already substantial tendency of top echelons to micromanage, it might also lead to more careful consideration of alternatives, a recognition of a greater need for flexibility of mind and habit, and constraint on the impulse to affix blame on the basis of a single instance. It also underscores the paradox that a doctrine, concept, or plan constitutes an axis of action and thought that a foe can recognize and capitalize on. Whatever advantages may be gained by keeping such indeterminacies in view, many aggressive-minded leaders, military and political, are likely to see such a weltanschauung as a form of "wimping out." In most military institutions, striking a balance between such ambiguity and the powerful forces running toward simplification and ordering will be very difficult, if not impossible.

THE FINAL FRACTAL FRONTIER: INDIVIDUAL BEHAVIOR IN WARFARE

Moving from that generalization to the more tortuous conceptual terrain of human behavior in war brings into view a virtual rainbow of behavioral variations among individuals at and between different levels. Complexities appear at every turn, for example, the differences among what is defined as proper and appropriate in selection and training, what is actually rewarded in peacetime, and what is needed in an operational setting. While Norman Dixon in *The Psychology of Military Incompetence* and many veterans have pondered over that discrepancy,[11] there are no clear patterns in that area. The wide diversity of individual behaviors in armed forces certainly belies the sense that military selection and training tends toward rigid uniformity. Military theorists and analysts have nevertheless tended to abstract human behavior, ignoring aberrant activity or viewing it with detachment. Military and civilian war games and combat models usually ignore panicked flight, straggling, willful desertion, and indiscipline. The ardent bellicosity of all participants is taken for granted, just as U.S. Army estimates of the situation always describe morale as "excellent." Those contrast, in theoretical terms, with breakpoint models based on one side in a battle giving way at a crucial moment, and in the practical realm, with operational plans and orders that designate straggler lines and control posts, and anticipate combat stress victims.

In the bureaucratic milieu of command-staff process, of course, individuals are described as aggregates, as numbers or standard phrases like "bodies" that convey little or no sense of their individuality although fiction, letters, diaries, and memoirs suggest how uneven rites of passage and military discipline are in actually standardizing behaviors, reflexes, and attitudes. The concepts of "bonding" and "cohesion" tacitly recognize such diversity, and ironically grant little weight to allegiances toward nation, community, or the armed forces' hierarchy relative to the small group, the prime mechanism whose social "glue" keeps warriors functioning in the chaos of battle. Those constructs also imply that not all those who go to war are equally ready to fight. Some paradoxes and inconsistencies have been left ill-defined or unaddressed, for example, the fact that conscripts, volunteers, and regulars have sometimes fought well and sometimes badly, as units, and as individuals, that each individual's performance tends to vary from situation to situation, and that leadership is not the same thing as commanding under the aegis of formal authority and discipline. The dissection of such military organizational dynamics has been an uneven practice, although many veterans tell of such anomalies as enlisted men and noncommissioned officers taking command when officers faltered, and how difficult it was amid the furor of battle to tell how well any particular person actually performed.

Measuring such dynamics and weighing evidence is also clouded by some claiming falsely to have been warriors and heroes, and some who served claiming to have done more than they did. Others found their urge to fight much diminished when actually engaged in battle, or fared better on some occasions and less well on others, like the main character in Stephen Crane's *The Red Badge of Courage.* While some suffered from having not lived to up to fictional or unattainable standards (labeled as the John Wayne syndrome in Vietnam), dispersion of warfare in time and space that resulted from technical advances made combat experience for many more a matter of enduring danger and tension than directly engaging a foe.[12] That is especially true of ground combat, which, even in its most static formats, is fluid, formless, and episodic, and those engaged frightened, excited, often tired, hungry, dirty, and physically miserable while enduring long intervals of boredom and apprehension between peaks of violence. In aerial combat, blurring relative speeds and wracking physical stresses limit perception, and in naval operations, the majority of ships' crews are occupied in tasks unrelated to the course of battle, locked in a function below decks that allows little or no direct sensing of specific danger or the course of operations.

Beyond the clouding of memory by time, the willingness to recall or relate stressful events tends to be inversely proportional to intensity of experience, that is, those more willing to talk may have less to say. Although most military historians and journalists have accepted such testimony with relatively little skepticism, there is virtually no way to authenticate it, aside from aerial gun-cameras. Individuals' recall of such events can no more than approximate what actually happens, and battle participants at best can recount no more than what they remember of what they endured.[13] Virtually no one at any echelon can validly relate every detail and exact sequence of a battle or campaign (the "big picture") in which they were involved.[14]

Although war is generally characterized as being generic, individuals' experiences of it are unique and have become even more so with the increase in subtypes of combat. As an example, air and naval warfare combatants often experience more personal comfort and less grinding fatigue than those on the ground. Such relative leisure, however, also provides opportunities to reflect, imagine and speculate, generating special kinds of stresses, and the repetitiveness of submarine–antisubmarine operations and air operations leads to some blurring and lumping together of separate experiences, especially the mission routine and anxiety-release cycles, like bombing missions, in which "grooving in" may overcome psychic defenses that might otherwise suppress the memory of specific traumas.

Further complexities arise at the level of individual action and behavior, including the varieties of motive, from bravado to deferring to coercion, that have led humans to war. Some have risked danger to prove themselves or to accede to social pressures, and have often found that experience misaligned with their anticipation of the conditions of battle and their own reaction to it. The impulse to fight is far from being universal, and willful and reflexive avoidance are well recognized in coercive conscription, military disciplinary codes, straggler lines, battle police, and brigs and stockades. A browse through Medal of Honor or Victoria Cross award citations will suggest how some heroes were not seekers after danger, but acted out of impulse. Some documentation lies in medical and court-martial records of those who discovered that it was an inner flaw that drove them toward action, or reached a shearing point, all within the sense of the popular phrase the "test of battle." The uncertainty of foretelling how someone will perform in battle has been dramatically visible by such cases as Audie Murphy, Alvin York, and Rodger Young, all mild in demeanor but lions in the field, while, to his own surprise, T.E. Lawrence, caught up in the tides of war, found it fascinating and compelling.

Like the phenomenon of the "unlikely hero," the genesis of human violence and aggressiveness remains perplexing, and to explore it is to enter a vast web of expanding detail toward the inner recesses of human consciousness, personality and, ultimately, nature. Tracing those perplexing psychic dendrites may be seen as something best left to clinicians, or dismissed as intractable, like the landscape background in a portrait, or static on a radio. Nevertheless, complexity theory underlines the potential for such small increments to yield disproportionately large results in war and presents historians and analysts with the dilemmas of trying to identify and weigh such evidence. Unfortunately, that process is as subjective as the individual perceptions upon which they base their portrayals of the past, and is further complicated by the interplay of sexuality, violence, and war, which is treated as a muted background theme by most military historians,[15] often anecdotally, or abstracted as medical statistics. While the widespread raping in the wars of Yugoslav devolution surprised many,[16] crude, brutal sexual language is nearly normative in armed forces,[17] and violence-linked sexuality is a pervasive subtheme in mythology, folklore, psychology, criminology, art and literature (e.g., T.E. Lawrence's long-banned *The Mint*, James Jones's *From Here to Eternity*, and Leon Uris's *Battle Cry*). On a parallel track, seduction and promiscuity are also virtual clichés in espionage literature, from the lurid reality of Mata Hari to the garish fancies of Ian Fleming's James Bond novels. Beyond that, clinicians have noted links between delinquent group and military behaviors relative to sexual psychopathy.[18]

Sexual drives are restricted, blocked, or channeled to some degree in virtually all human groups and institutions, but especially so in gangs, religious orders, certain types of schools, jails and prisons, and military organizations.[19] There has been little direct measurement of the function of sexual isolation in armed forces as a generator of aggressiveness, but major, lengthy wars, along with their mass mobilization and displacement, have been marked by surges in basic, violent emotions. In World War II, the massive venereal disease epidemic in the United States and Britain reached major crisis proportions; it was brought under control by the introduction of antibiotics. Again, such dimensions of war are usually omitted from military history, or treated as marginal or of little or no consequence, which is not the case with war fiction.

Another disparity in individual attitudes and behaviors is visible in the range of attitudes of military professionals toward war and violence, and complexity and order. For example, Douglas MacArthur oscillated between pugnacity and hesitancy throughout his career. After Hiroshima, he declared that war as an institution was obsolete, then sought to expand the Korean War in East Asia in 1951, and finally, after being relieved, cautioned three presidents over the next decade not to commit major forces in that region. A considerable number of major modern commanders who displayed pugnacity on the battlefield later spoke out against war, including Allenby, Butler, Gavin, and Shoup, while Eisenhower titled his presidential memoirs *Waging Peace*. On the other hand, some, including Ludendorff, Patton, Radford, and LeMay, remained stridently bellicose. The pacific rhetoric of the former was not aberrant, but reflected the peculiar tension within the role of officers as overseers and wielders of violence in the name of the state. Violent posturing has tended to be a function of stratification, with noncommissioned officers and junior officers, who are more often engaged in actual fighting, usually expected to be more pugnacious in demeanor and attitude than their seniors. In keeping with the classical stereotype of *miles gloriosus,* fictional NCOs have often been portrayed as ardently violent in combat,[20] in their imposing discipline, and when off-duty, even if otherwise technically competent and self-contained.

In contrast, from the mid-1600s to the early twentieth century, field-grade and general officers were seen as responsible, self-controlled agents of higher authority, usually but not always brave, rarely bellicose, and at their best, balanced, firm in judgment—gentlemanly warriors who observed the laws and customs of war and restrained subordinates from excesses. Such ideals crumbled, both in fiction and in reality, in World War I, partly because of the physical separation of generals from fighting troops. The erosion of the role of officers-as-gentlemen, paralleled by widespread abandonment of laws and customs of war, was reflected from the 1930s onward in films depicting friction between officers of different social classes.[21] That transformation was also reflected in the emergence of several dictators from NCO barracks and from officer corps drawn from underclasses.[22] A decline of chivalric values was also reflected in an epidemic of atrocities, including some that were ordained at the level of high policy, and an end to formal standards of refinement in officer's messes.

Such social turbulence in the military domain was compounded by the diversification of roles due to the division of labor stemming from technical advances. From the mid-eighteenth century onward, aside from some types of surface naval warfare and strategic bombing, an ongoing dispersion of subelements was forced by increases in accuracy, velocity, volume, explosive power, and range of weapons. Although that change in rates required independent judgment and initiative, many old formats were retained, sometimes at high cost, as in the slaughter of close-order tactical formations in the Boer, World, and Korean wars. Rigid uniformity in dress and linear formations dominated entry training and rites of passage, even though such strictures were assailed before the Napoleonic Wars by such innovators as Marshal de Saxe. Eroded during those conflicts, the old formats reemerged after the Napoleonic Wars, defended on the basis of tradition, but partly symptomatic of the reimposition of social hierarchies, especially officer-enlisted tiers of rank, ceremonies, costumes, organizational tiers, terminology, and other practices and attitudes such as saluting. In spite of the weakening of many "old school" social practices, linear and symmetrical tactical formats persisted in Western nations' ways of war, for example, the bomber fleets of the Eighth Air Force of 1942–45, the

Falklands "gun-line," and the elegant symmetrical gestalts of DESERT STORM.

Curiously, that dialogue between complexity and order has not been examined to the extent warranted by either its historical or functional import.[23] Few within armed forces, let alone military historians, analysts, or critics, have questioned how such practices and structures have actually enhanced or hindered military performance. A most critical question relative to complexity is how much they reduced or augmented complexity, or led to chaos. At the level of individual behavior, some forms and styles have held over a long time, like the sangfroid of British Guards and French Foreign Legion officers, and the thrustful, tough mien of U.S. Marines. The angle of divergence between such attitudes and practices in armed forces and their parent societies remains more vaguely sensed than precisely plotted, in spite of the resultant turbulence, from the widespread mutinies in World War I to the "fraggings" (assassinations) of officers and NCOs in the U.S. Army and Marine Corps at the end of the Vietnam War. Although military theories on such matters have originated at points around the political compass,[24] some undercurrents in military organization and method have continued to run toward the old formats, and a valuing of order for its own sake. In spite of egalitarian rhetoric and experimentation in their early phase of formation, revolutionary armies tended to retain, refine, and strengthen traditional forms, as did the English "New Model" in the mid-seventeenth century, the armies of the French Revolution in the late eighteenth and the armies of Communist Russia and China in the twentieth. As with other complex social phenomena, charting such organizational dynamics is hampered by inertia and impressionism, and lies far from clinicalism or science.

TURBULENCE AT THE TOP: COMPLEXITY IN HIGH ECHELONS

Individual variances that feed complexity at lower levels of military and naval hierarchies take on a different dimensional cast when viewed from the peak of such pyramids, inasmuch as strengths, weaknesses, quirks, whims, and the like may be dramatically amplified or wholly obscured by the mechanisms of power. By the end of the American Civil War, or earlier, some might argue, orchestrating vast arrays of forces in war had become impersonal and imprecise. Although from that point on, much of the business of command-and-control lay in the hands of large numbers of anonymous staff officers and commanders, the mystique of a heroic martial leader in absolute control as maestro remained the capstone of the military ethos. That was visible on both sides in the Gulf War of 1991, and in the fact that all major twentieth-century dictators, and many minor ones, excepting Mao, wore military garb, although few of them actually commanded major forces in war.[25]

Another source of complexity and chaos in respect to major leaders is their state of health, often misrepresented, but which sometimes substantially influenced events. In World War II, Roosevelt, Churchill, and Hitler suffered from major illnesses, as did most key leaders in the Suez Crisis of 1956, and both Lyndon Johnson and Ho Chi Minh at the peak of American involvement in Vietnam.[26] Although many such examples of individual aberrance or pathology had come into view by the time the cold war intensified, and as nuclear war battle management networks were emplaced, the dilemma was addressed only peripherally. Elaborate systems were instituted for vetting military controllers in the nuclear chain of command at lower levels, and public concern about the "red phone" mounted after President Eisenhower's various health crises, and surfaced again in the presidential election campaign of 1972. In the 1950s, the appearance of several fictional "doomsday" scenarios was paralleled by a change in the line of succession in a military emergency. The creation of the National Command Authority (NCA) and a constitutional amendment were both aimed at clarifying the problem of succession in ambiguous situations. Those arrangements, however, did not seem to work quite right when Eisenhower underwent major health crises in the mid to late 1950s, as Nixon became beleaguered during the Watergate affair in 1973–74, or when Reagan was shot in 1981.

Throughout that era the "red phone" (which never existed as such) symbolized the nuclear-age role of U.S. presidents as crisis-copers and, potentially, as nuclear war battle managers. In spite of the fact that most presidents since Roosevelt suffered major health problems, the assumption that major leaders were essentially sound and rational persisted in both the structure and theories of power.[27] Considering those factors in light of complexity theory offers a special perspective, especially in respect to small influences causing unexpectedly great mischief.

THE DEEPER ROOTS OF WAR'S CHAOS

Although Hitler's sanity and that of other Nazi leaders was widely doubted, the victors treated many surviving Axis principals as criminals, including the clearly demented Rudolf Hess and Julius Streicher. The detailing of their atrocities, from the Malmedy Massacre and the Bataan Death March to the Holocaust, obscured the fact that some Allied nations' policies and excesses contravened international law, the customs of war, or their own prewar diplomatic positions, especially unrestricted submarine warfare, aerial bombing of cities, and the use of napalm and white phosphorous. Beyond that, the rhetoric of major

Allied war leaders frequently matched the substance of such acts. In 1941, for example, Churchill, in a speech at the Guildhall, rejected the idea of limiting the effects of bombing on civilians and vowed massive retaliatory bombing of Germany. Roosevelt's request to Congress for a declaration of war following Pearl Harbor was no less wrathful, if less specific about method. Perhaps most bellicose of all was Air Marshal Harris, chief of the Royal Air Force's Bomber Command, who broadcast threats to the civilian population of Germany.

However valid Freud's view of war's meshing of lofty moral purpose with deep drives toward viciousness, those polemics added layers of complexity to that conflict as the Allied leaders cast themselves as war chiefs as well as defining postwar goals and aims, well away from Wilson's striving to be an aloof peacemaker. It is not possible to measure the exact congruence between their exhortations, the amplification and diffusion of them by media, and their effect on friends and foes, especially as a stimulant to the determination of Axis forces to fight more ardently. At the very least, those postures lay well away from the Clausewitzian model of war as the pursuit of policy goals by violent means, and perhaps closer to a collective tantrum. Whether they reflected general ethical and moral decline, or attempts to enhance the fighting spirit of nations imbued with pacifist sentiment, they added to the complexity of that war and created new views of war, some of which have not yet faded away or been fully traced in their effects.

That is not surprising, since attempting to appraise such intricate forces and the effect of the personal quirks and impulses of both leaders and followers on warfare, including elites' secret interactions, leads down yet another widening fan of causal corridors, which, like small turns of fate, accident, weather, and catalysis, can only be estimated, if they can be imagined at all. Many sensed before Freud that the domain of the nonconscious or subconscious and its influence on action usually lies out of sight, even to the actors themselves. With only fragments of testimony and recorded behavior beyond formal documents offering insight into the state of mind of leaders or "lower participants," there is no way to fully plumb their deliberations and actions. The dynamics of complex social interaction, large and small, including warfare, cannot be assayed, nor can the collective mood of societies at large, beyond a general sensing based on documents, films, tapes, art, and so forth, or public opinion polls.

As difficult as that problem is in political, diplomatic, or business history, it presents analysts of war with a further obstacle. As noted earlier, most military documents are stylistically flat and impersonal, shaped by many stages of pruning and smoothing in staffs and bureaucracies. Considering what goes on in war, such communications reflect little of the nature of the business at hand, or of the emotions generated by deliberately attempting to kill, injure, and terrorize, or incur the fear of violence. Most official documents are arid and sterile and rarely include references to what the physical results will be of a policy, doctrine, plan, or order couched in rational logic and terms, code-named, and set forth in bland semantics, but aimed at wreaking mayhem. In contrast with wartime political polemic, grandiose orders of the day, or lurid recruiting posters, little in the official "paper trail" reflects the great churning stew of emotions that causes societies to mobilize to inflict and risk mass violence and destruction.

At the furthest reaches of individual human motive and behavior lies the deepest dilemma raised by complexity and chaos theoretics: whether some inner impulses that lead a society to war, and policies, judgments and acts, both of leaders and led, without their awareness, are aimed at some private, nonconscious agenda, such as the seeking of death, pain, defeat, or failure, or the venting of blind, normally suppressed rage, rather than the conscious desire to damage, kill, and attain victory. That range of impulses cannot be dealt with beyond deduction and uncertain conclusion,[28] both because of inadequate instrumentation and methodology, and because of lack of agreement on the validity of historical analyses based on attempts to probe such psychological dimensions. Yet that does not mean that such forces have had no significant effect on war dynamics, or that similar problems do not affect events at the lower end of the chain of command as well. Here, again, evidence of what happens at the "business end" is filtered by the command-staff process as it imposes matrices of order on turmoil. We can, therefore, only speculate how much rationalistic structure and procedure have affected commanders versus how much they have bent it to their purposes, or to what extent their deviating from operational norms and procedures, or from the course of rationality, yielded success or failure.

As with opera prima donnas, the eccentric behavior of many military leaders has often been seen as lying within a range of normality for people operating under great stress. Hence such terms as "command discretion" or latitude. Again, it is very hard to determine if quirkiness or unorthodox actions in a specific case were deliberate or born of fear, anxiety, inspiration—or pathology. As suggested earlier, such behavior has often been attributed to creativity, insight, and genius, and odd behavior has been seen as evidence of mild eccentricity or reaction to stress. While there is no equivalent to the "care of the flier" program for commanders, incidental monitoring by subordinates, periodic medical examinations, and inspector-general and law enforcement systems serve as constraints on commanders' behavioral latitude. Although there are no firm, objective indices, mental and professional competence are generally assumed, and relief for cause is relatively rare.

Beyond the question of physical health lies the sanity and competence of elites, a major theme in drama and in the history of imperial and royal dynasties that has persisted into the age of democracies and dictatorships. Root causes of apparently nonrational behavior on the part of commanders have rarely been traced. Well-known cases of careers marked by cycles of triumph and disaster include Napoleon, Lee, MacArthur, and Churchill. Were those apparent cycles products of chance? It was not a blind turn of fate that MacArthur failed to follow through on his order just before Pearl Harbor to disperse the heavy bombers from Clark Field, or that led him to abandon his army's vast food reserves as it withdrew into the Bataan peninsula. His loyalty to his arrogant and unevenly effective intelligence officer, Charles Willoughby, yielded several close calls and finally led to the misestimation of Chinese intervention in Korea that marked his final descent from glory.

Such dimensions may be seen as ephemeral and impossible to chart, but they are vital nonetheless, since military historians and journalists have usually attributed the outcome of military operations to commanders' decisions and actions. Many commanders took great risks to the point of recklessness and won victory, but leaders' performance cannot be set on a scale of performance, since there is no way to know what would have happened in alternative circumstances. Judging that by the outcome offers little help because in war, in the simplest terms, one side wins and one loses. Paradoxically, not all winners have been judged to have been professionally competent, nor all losers incompetent. Military historians and journalists have severely faulted or slighted some commanders who never lost, most notably Marlborough, Wellington, Grant, Haig, Montgomery, and Eisenhower. While commanders have sometimes fled, broken down, or rejected what seemed to be good advice, such unambiguous cases have been relatively rare. Critics often overlook dimensions such as context and momentum, and pressures to act quickly—or precipitously—whether for tactical, strategic, and political reasons, or to gratify the anxieties of those hungering for action for its own sake. From the standpoint of complexity, judging commanders as virtuoso decision-makers and tacticians is blurred by the myriad vortexes and crosscurrents of warfare, as is the pervasive model of armed forces as an integrated organizational pyramid, responsive to deft direction. That may have been the case with Dan Morgan at Cowpens, or Stonewall Jackson in the Shenandoah Valley, but it is not always so easy to plot the crucial points at which a commander influenced the flow of a battle.

CYBERNETICS AND COMPLEXITY: THE SOVIETS AND NONLINEARITY

As ongoing technical advances in communications technology increased the rate, volume, and distance of information transmission over the last century and a half, armed forces were presented with the recurrent paradox that progress failed to yield proportional improvement in the exercise of command. Especially important from the standpoint of complexity is the way that overload caused by those advances hampered attempts to perceive patterns and trends amid the increasing deluge of data emanating from many not wholly orchestrated sources. In the Soviet Union, ironically, in view of the Russian preeminence in nonlinear research, the resistance of Party and military leaders to the cybernetic revolution was not overcome until the 1960s. While many command-and-control problems, including information glut (*potok,* or flood) were explored rigorously by Soviet scientists and military analysts,[29] it is not clear how much the long-standing interest in nonlinearity was brought to bear in that research, or how much those explorations altered their view of a collision of simple, major force vectors in the dialectic model lying at the heart of Marxist-Leninist philosophy.[30] It is also hard to say how much it threw light on the high costs in material and human suffering, both in war and peace, that the Communist regime imposed in trying to monitor and control increasingly complex and evolving systems.[31]

In World War II, as the Red Army's mobility and communications increased, allowing greater flexibility and range, its tactics became increasingly fluid. From 1943 to 1945, it assailed the Germans with phased combinations of force, its preparations well screened by elaborate *maskirovka* schemes, arraying elements under a logic of "correlations of force" that smothered the Germans' maneuver-based tactics which had been so successful in the opening campaigns of the war. Aside from Stalingrad, the Red Army did not attain the quintessence of operational art, the battle of encirclement, that the Wehrmacht had carried out repeatedly during the first few weeks following the German invasion in 1941. The Soviet organs of state security were wielded heavily to counter disintegration and chaos, in some situations using exemplary executions to discourage headlong flight. After that great retreat, NKVD units were salted among Red Army elements as stiffening forces, a proverbial stick as counterpoint to the carrots of exhortations and images and the invocation of pre-Bolshevik-Soviet patriotic history. Another such attempt to divert a dialectic vector was seen in the use of the state security elements in the late 1920s and '30s, when the Soviet regime had also massively underestimated the scale and complexity of social forces they sought to control. In 1941, as Stalin, in near-despair, ordered drastic de-Communization of the Russian war effort to engender patriotic spirit against the Nazis, many Russians witnessed the failure of the regime's rigid linear planning and the faltering of its central control mechanism, and the subsequent triumph, in uneven stages, of expedients, improvisation, intuition, and nonrational forces. The slogans, formats, and dicta of the 1920s and '30s were discarded amid the collision of a virtual blizzard of vectors in war that led to recombinations, realignments, and the generating of yet more vectors, implicitly calling into question the validity of the dialectic.

Not surprisingly, the shapers of Soviet military doctrine, with their catechistic footings in Marx, had cast out Clausewitz, even though both Lenin and Mao found his work of some interest. In Communist epistemological terms, Clausewitz, like Marx a generation later, was caught in the dialectic collision of the "mechanistic materialism of the Scientific Revolution and the Enlightenment"[32] and the turbulent metaphysical waves of Hegelian idealism and Romanticism. Alan Beyerchen has pointed out how, as Clausewitz strove to construct a rational framework, he was pulled away from classical linearity toward that threshold of turbulence that novelists from Stendhal and Tolstoi to Mailer crossed more boldly, visualizing wars as collisions of great, indeterminate forces that, like gas clouds, could not be fully sensed or precisely orchestrated, and were more a confluence of massive social forces than the purposive enterprise portrayed in so much of military history.[33] Clausewitz's ambiguities and his deferring to chance stood sharply opposed to Marxist-Leninist-Engelsian faith in the identifiability of the laws of history. If Soviet cold war strategic doctrine ever comes fully into view, we may be able to see how it enfolded the colossal chaos implicit in nuclear war-fighting, and if, for example, the massive throw-weight of the Soviets' largest ICBM was seen as a military-strategic dialectic vector aimed at countering nuance with mass, in keeping with their concept of the "correlation of forces" as a variant of dialectics.

COMPLEXITY OUT OF SIMPLICITY: COLD WAR PATTERNS

Especially perplexing is the frequency with which apparent trends toward simplification actually produced increasing complexity in the braided stream of history. Immediately after Hiroshima, for example, many, including senior generals like Arnold and MacArthur, saw the atomic bomb as having brought warfare to a virtual dead end, but it soon became clear that fear of using nuclear weapons was merely squeezing international conflict into new and diverse formats. Superpower rivalry, like opposing magnets' effect on iron filings, produced sprays of branches around and between the opposing poles, further branchings of which led to a widening range of conflicts, some of which remained hypothetical, including general, all-out, and theater, and some of which were real, such as regional, local, limited, and small wars, "low-intensity conflict," and "wars of national liberation": a range of complexity fed by a technical competition that generated ever greater numbers and types of "weapons systems." As one index of that evolution, at the end of World War II, U.S. Army ground forces deployed about fifteen such "systems," but by the late 1980s, there were about fifty, of far greater technical sophistication, constituting a fractal range whose branches extended well beyond military boundaries in to social and economic systems around the world, as they absorbed material and technical resources.

During much of the cold war, the Western bloc relied on nuclear weapons and technical sophistication to deter the numerically superior foes of the Eastern Bloc. Neither sides' full advantage could be brought to bear in the "hot" wars of the cold war because of the distortion fields generated by their nuclear arrays. Thus the French in Indochina, the British in Malaya, and both, as allies in the Suez Affair, the Russians in Afghanistan, and the United States in Korea and Vietnam were all constrained by anxiety over those conflicts getting out of control and leading to a superpower nuclear exchange. From Korea to the Gulf War, above and below the threshold of actual combat, the Western powers were frustrated by camouflage, deception, and labor-intensive systems that sometimes turned their technical advantage into a propaganda debit. At the highest levels of strategic concern, the complexities and nuances of such "limited wars" remained secondary to the monolithic contingency of Eastern and Central European *Hauptschlacht*, although some feared that such marginal conflicts might, through "catalysis," trigger a "general war."

As a result, many such cold war "sideshows" were fought by forces initially trained and equipped to engage in sustained, intensive "conventional" warfare. Beyond consuming and diverting enormous resources, those skeins of power created vast fields of potential, like charged batteries and capacitors. In the last stages of the cold war, many analysts saw that complexity as pregnant with true chaos, whether the superpowers went to war deliberately or were drawn into it by accident or catalysis. Anxiety eased considerably as the cold war evaporated at the end of the 1980s, even though it soon became clear that the major Western powers' involvement in remote military operations might not decline in the new order—or disorder—of things.

THE CRAZING OF THE *GANZFELD*: COMPLEXITIES OF THE "NEW WORLD ORDER"

As the grand dialectic of the cold war faded, perplexing subwhorls of complexity emerged, presenting analysts, policyshapers, and leaders with an increase in critical tensions and war contingencies. Although actual "small wars" did not increase markedly, they were no longer obscured by the glare of superpower-related tension. More critically, it was not clear if misalignment and drift in the new *Ganzfeld* might lead to the emergence of aggressive regional powers or coalitions. The slide toward chaos in the "new order," blurring boundaries between war and peace that Martin Van Creveld and others have noted, has made it harder to tell where "pure" warfare ends and "low-intensity conflict" begins.[34] When examined through the lenses of complexity, that smudging of war-peace boundary states makes war dynamics seem more a "fuzzy" process than a well-demarcated set of events. As a result, considering international security from that standpoint presents a challenge in determining whether—or when—"small wars" around the globe as an aggregate are merely "simmering" or approaching a state change, like clusters of bubbles and turbulence in a heated pot that appear before general boiling erupts. Related questions are whether that will happen if certain levels of heating are maintained, increased, or varied, or suddenly increased or decreased, and if it is possible to construct indices of those state conditions or control them.

THE QUESTION OF DIFFUSION

It seems unlikely that complexity theory will, as a kind of conceptual gunpowder, breach or topple such solid bastions of cultural attitude and practice as the hunger for order, categories, rules, firm and clear judgments, simplification, indicators, or political systems based on clear contrasts and conflicts. Considering that, appropriately enough, generates some paradoxes. If basing such prognosticating on sketchy evidence is directly antithetical to the implications of complexity theory, it lies closer to the essence of creativity theory. If complexity theory ultimately has substantial effect outside the sciences, it would be in keeping with the logic of small influences having major consequences. Perhaps heightened awareness of complexity phenomenology will reshape perspectives on sorting details and labeling information in various professions and disciplines, by throwing a less pejorative light on terms like as "trivial," "spurious," and "marginal"—and on "chaos" itself. If it weakened the impulse to readily discard seemingly slight or exotic influences, it might also increase or reduce complexity per se—or lead to chaos. In the present political environment in most countries, few leaders or analysts are likely to be inclined to confess "predictive hopelessness,"[35] that is, admit that matters are out of control or beyond their comprehension, even when they obviously are. Such posturings of confidence under stress is a pancultural mythos, and steadfastness in the face of disaster, from Horatius at the bridge to Robert E. Lee and Winston Churchill, has been awarded special mana. Contrarily, there is also a general appreciation of the pervasiveness of confusion and uncertainty in complex organizations and in warfare (e.g., the popular expressions SNAFU, "fuzzy sets," and FUBAR). Such ambiguities have been popularized from time to time, in formats ranging from *Omni* to *Scientific American,* educational television, and best-selling books such as James Gleick's *Chaos.* They also entered the realm of popular culture, most notably in Michael Crichton's novel *Jurassic Park* and Steven Spielberg's film version of it, and fractal software well known to the computer-familiar "Generation X."

Perhaps a successor to Marshall McLuhan will explain how spatially unlinked and gossamer gestalts of radio, television, and personal computer networking provided the tolerance for divergence and sensitivity to subtle fusions in complexity, further undercutting the rigid centralism imposed on much of the world through the early modern era by mercantilist imperialism, followed in the nineteenth century by webs of railway, steamship, telegraph, telephone, and electrical power networks. Much of that was underpinned by symmetries detected by scientists such as Newton, Descartes, and Napier, and later beaten into philosophical swords, most notably by Marx, who visualized revolutions as the "locomotives of history," and who, with Engels, followed Hegel's recasting of Newtonian physics into laws of motion to explain economics and the flow of history.[36] Small wonder that the Bolsheviks were enthralled with Frederick W. Taylor's methods of efficiency and industrial engineering, but not all were so taken with those linear patterns of progress. Some, including Marxist deviationists, steered against the rationalistic centralism that was commandeered by a host of dictators, from Cromwell to post–cold war caudillos. The more openly hostile included the economic theorists von Mises and Hayek, who exceeded Adam Smith in challenging the feasibility of imposing order on market and social dynamics, which they envisioned as generating intricate dynamics of perfection, like Newton's harmony of the spheres which Einstein and other theoretical physicists set awobbling.

As such concepts as complexity, chaos, nonlinearity, and fuzziness diffuse beyond the borders of science, mathematics, engineering, and economics, some distortion must take place in the conveying of such concepts to wider audiences. For researchers and consumers alike, each one coming to the conceptual well has a unique focus of interest and ability to comprehend the technical dimensions, but the former and only some of the latter will rely on equations. Since many cannot comprehend them, or are put off by the elaborate jargon suffusing such research, it is likely that, as with cybernetics and General Systems Theory, the drawing of broad analogies and effects may match or exceed the conveying of detailed specifics. Leaving such mathematical apparatus behind produces gaps in understanding, and less precise definitions, but may no more inhibit the diffusion of the general sense of such metaconcepts than it did those of Newton, Darwin, Freud, Marx, or Hitler, and of many religions and cults. In such cases, very fragmentary or superficial understanding of central tenets did not prevent uneven and distorted radiations from them having very powerful effects on thought and action. That potential distortion presents researchers and journalists with the burden that all translators bear.

In any case, complexity and chaos, like Lorenz's butterfly-wing metaphor, fractals, and fuzziness, have gained substantial popular currency.[37] It is not clear whether the rising awareness of the disproportionate influence of small, incidental, or random factors constitutes a significant "paradigm shift." It does move away from simplistic metaphors like the Newtonian-Hegelian-Marxist collisions of vectors to a neo- or transdialectic in which historical processes are envisioned less as a collision of locomotives,[38] but more like the mingling of the inhabitants of adjacent beehives mingling in an earthquake. Although many complexity-chaos theory elements have long been in view, it is not clear how much the marginality of such concepts was a function of a kind of incubation, reflexive rejection of their essence, or due to the lack of adequate calculation aids. Factors other than the increased computer capacity that made chaos-complexity more tractable to portrayal and analysis may have led to acceptance of their feasibility. Over the last century, wars, crises, political disruptions, and natural catastrophes, graphically conveyed to an increasing portion of the world's population by mass media, repeatedly demonstrated a chronic inability of elites and experts to anticipate or deal with complex phenomena, in contrast with the posturing, symbols, and rhetoric of intuitive power and infallibility suffusing the upper tiers of hierarchies.

What are the likely longer-range consequences of those trends? Nonlinear concepts blur the precise lines and points that many historians use to mark the ranges of epochs, eras, and periods, or to trace trends and tendencies, while further eroding their sense of surety in ascribing causation. As these constructs, along with events, add to the sense of a decline or social disorder, will that lead to denial, or to political or religious counterconcepts? Chaos and complexity theoretics offer some explanation why leaders, civilian and military, have so often failed to understand and control turbulence to anywhere near the degree that they usually claim a special capacity to do. The implications regarding power resemble those Classical, Renaissance, and Enlightenment models that shapers of the American Revolution and the Constitution had in view when they sought to limit what they saw as a universal, sinister lust for power. They had a particularly good vantage point to watch elitist corruption, greed, and ineptitude generate chaotic policy in peace and then war. Although the fallibilities of power have been long recognized in popular culture and common sense, as in such images as "the Emperor's new clothes," the growing visibility of elite incapacity has been placed over much of the world by an increasing use of vicious methods, from state-sanctioned torture to insurgent terrorism, all aimed at inducing chaos.

THE TEMPTATIONS OF "WHAT-IFFING"

Causal dynamics and "what-iffing" have taken such diverse forms as so-called bull sessions, university discussion sections, war gaming, and the "alternative futures/histories" of fantasy and science fiction.[39] Less academic and slightly less fanciful variants of "what-iffing" are depictions of alternative futures based on extrapolations from relatively solid data, such as General Sir John Hackett's *The Third World War,* and the novels of Tom Clancy, Larry Bond, and Michael Crichton. Such practices are discouraged in the academic training of historians, which, whether based on presumptions of coherent causal patterning, or the uniqueness and inevitability of events, assumes their reducibility to reasonable description and explanation. The closest that academic historians approach such speculation is in framing loose questions such as the following: "Did incipient corruption and maladministration in the Nazi government, or specific policy or strategy choices cause the Third Reich to lose the advantage gained by its early military victories?"

At first glance, it is easy to see why "what-iffing" is unpopular among historians. In a world of limited resources, it makes sense to focus on facts, not visions or fancies. If an event happened because it was the only thing that could have happened, generating alternatives is a waste of time. Such distracting fantasies may distort the gathering and analysis of evidence, and consequently prevent describing a reasonable approximation of events, tracing of causes, and ascribing of meaning. On the other hand, "what-iffing" is an exercise that students of complexity may find more interesting and useful. In military history, that would involve probing the dynamics of warfare, including the perplexing anatomy of victory-and-defeat, to a degree of depth and precision previously beyond the capacity of most if not all individuals, as increasing computer capacity augments the arraying of data, calculating of complexity, and identifying of the transition from complexity to chaos. Presuming that happens, complexity and chaos theories will present a challenge to the tone of confidence flavoring historical expositions that often beg the question in identifying causes and draws meaning that are out of phase with evidence, or that are based on evidence arrayed to support assumptions or parochial agendas. That will become more apparent as complexity-chaos theories alter perspectives on how much and what kind of evidence is required versus that previously viewed as relevant or adequate.

Before the appearance of such theories, some historians questioned whether the course of history is like a river, shaped by smoothing forces, laws or purposive currents, as humanity moves along a course of destiny unaffected by minor influences. Recent assaults on such "covering laws" are Donald McCloskey's challenge to the model of norming tides in history,[40] and John Lewis Gaddis's assertion that "we tend to bias our historical and . . . theoretical analyses too much toward continuity."[41] When that logic is applied to military history, it raises the question whether there is any sure way to sort out which events in war and battle have been or might be those "small changes . . . [that] may occasion radical changes in their output. . . ."[42] While some recognition of that dilemma is visible in revisions and changing interpretations in academic historical discourse as fresh data or insights generate new perspectives, it is not clear that sufficient evidence can ever be amassed to support the tone or claims of certainty so often found in historical explanation. In military, political, diplomatic, and business history, that problem is reflected in critiques by "insiders" who were involved in events described by historians, and by journalists who were not. Such a sense of exclusive knowledge is often reinforced by the informal and secret procedures that thwart monitoring by foreign and domestic political foes in democracies and authoritarian systems alike, and by a sense of social exclusivity. Such claims of special perspective and insight degree, although relatively valid, tend to overlook how those constraints and the limits of individual perception deny the gaining of a purely objective and fully detailed view from any direction, and absolute surety that what they know was all there was to know, or that it was as significant a portion of causation as they presume. Beyond that, secrecy creates a vacuum that leaves the shaping of history to such sculpting forces as speculation, fabrication, hagiography, and polemics.

THE GULF WAR: CURSORY OBSERVATIONS FROM A NONLINEAR ASPECT

By the mid-1990s, the expanding torrent of revelations of the Gulf War offered some perspectives on questions of complexity, especially since the conflict seemed far simpler and more clear-cut in format than the Vietnam War. Although many pundits were proven wrong in the anticipation of a bloody shambles, some observers and analysts, perhaps most notably Trevor DuPuy and his associates, who were intimately familiar with U.S military professional training and style, anticipated the shape and course of operations with a high degree of precision.[43] As Operation DESERT STORM unfolded, television images of well-defined tactical formats and orchestration conveyed a pristine symmetry. Censorship kept much but not all of the administrative and logistical confusion out of view during the six-months' assemblage of half a million troops and much of the Allies' air forces, an effort that, although subsidized, heavily strained declining Western military resources. As in previous wars, much detailing was denied journalists, consequently hampering the view of complexity and chaos, an effect multiplied by the fact that much documentation existed only in the form of electronic data bits, and only some of that was reduced to hard copy.

The sense of coherence was also augmented by the schematically straightforward battle plan, basically a reversal of the feint–right hook schema of the British Eighth Army's 1942 attack at El Alamein, including an introductory air campaign, an amphibious feint, and a massive preparatory bombardment. While tactical control was marked by such linear formatting as the air support controllers' "kill boxes," it was not clear if reliance on personal computers, networks, and various software led military planners, commanders, and controllers to rely on such rectilinear formats, or if that was a generic property of desert warfare. Nor was it possible to determine how much the systems sensitivity and nonlinear mind-sets of many of the Allied combatants, as children of the "computer age," affected the flow of operations. Paradoxically, there was little evidence of the freewheeling maneuver warfare and emphasis on lower-echelon command initiative that became the idée fixe of many U.S. Army officers in the post-Vietnam era.

The descriptive limits of statistics and narrative descriptions leave it unclear as to how much the nonlinear potential of air power to break free of linearity through arraying and phasing varying sizes and configurations was tapped, or if it was constrained by rectilinear formats and plans. While Allied operations were aided by the rapid disintegration and flight of Iraq's air forces, the marginal effects of its defenses, and its technical inferiority, the Iraqis' widespread use of camouflage, concealment, and deception also raised some doubts about the measurement of effects, as had previous Chinese, North Korean, and North Vietnamese stratagems, also devised to offset mass, linearity, and technical advantage. Beyond the difficulty of determining how much of what was seen was real lies the larger uncertainty of whether the Gulf War was unique, or if it marked a line of evolutionary continuity and entry into a new dimension in warfare.

THE FUZZY ROAD AHEAD

After the Gulf War, the widespread sense of relative tranquillity and of a "new world order" being at hand that followed the demolition of the Berlin Wall faded quickly. In the wake of that conflict Western leaders and policyshapers faced fresh uncertainties as the fragmenting of former power complexes and shifts in various regions generated crises and wars. While the cold war's firmament of friction points had been more intricate than many chose to remember, there was clearly a growing sense of chaos in international affairs. Complexity theory helps to explain that turn of events, but offers little comfort, especially considering that the increase in the "n" of such friction points and boundaries, in Richardsonian terms, raised the likelihood that seemingly minor, subtle events and decisions might yield sudden, major consequences. One derivative paradox was that the easing of anxieties devolving from the ongoing superpower nuclear confrontation, and especially the dimensions of escalation and catalysis, were balanced by mounting concerns about "horizontal proliferation" in various regions. Another was that increasing complexity in a depolarized world undercut the capacity of all nations to monitor or move to deal with resultant turbulence, and looked very much like chaos.

The diffusion of bits and pieces of complexity-chaos theory, including simplicisms and distortions through the culture, may affect the diplomacy and defense policy of the United States and other nations, if only in offering a view why things so often go askew. To the extent that both such a worldview and trends in world affairs move away from focusing expectations and clear-cut polarities, that conforms with von Moltke's view of strategy as a series of ad hoc expediencies. As a result, American elites are presented with a special dilemma. Such ambiguity and uncertainty stand in sharp tension with Americans' proverbial hunger for clear definitions and goals, and their literal itch to resolve identified problems and complete tasks quickly. How much the mind-sets of most American and many Western elites have been shaped by logics of clear dichotomies and unambiguous resolution is visible in language, the formats of sports and games, and the education and training of military professionals, lawyers, engineers, and academics. Regardless of external realities, those impulses, along with the American tendency to couch foreign policy in moral terms, will stand in tension with at least some of the implications of nonlinearity.

That, of course, is not universally true, since some tides were turning before the surge in interest in complexity-chaos. Despite the pervasive influence of statistical mechanics, some engineers and scientists sought constructs and patterns to deal with ambiguous realities,[44] as did some military professionals and intellectuals.[45] Nevertheless, American leaders have tended to focus on close-range problems versus scanning, rely on linear projections of trends, and prefer simplicity to complexity.[46] Although nonlinear concepts have been entrained with such defense-related tasks as cryptography, cryptanalysis, and communications routing,[47] and fuzzy and rough sets have been applied with substantial effect in the realm of computers,[48] those efforts have lain in compartments of function, well away from the processes of power.

ON THE UTILITY OF CHAOS-COMPLEXITY THEORETICS

The ideas referred to in this study, and the many others that constitute the corpus of complexity-chaos-nonlinear theoretics, like all models, are at best imperfect latticeworks of thought. Humans cannot know full reality or describe it,[49] since only exact, full reality *is* reality, and any description of it is therefore abstract and approximate. The limits of human perception and of electromechanical extensions of it further blur the view of chaos-complexity, and add to it, allowing only imperfect, brief impressions of a relentless procession of instants of reality to be captured and stored in the disjointed crannies and bins of memory. Obviously, some attempts to gain a clearer view surpass others. While complexity-chaos theories offer fresh perspectives, are they really a better set of lenses, or only another perspective or variant of an ultimately intractable level of distortion? Since any valuation must be relative to other models, of what use are they to practitioners or analysts in any field, let alone in the violent and near-chaotic realm of warfare? How might sensitization to complexity-chaos affect leaders' and policyshapers' perceptions, plans, or actions, for good or ill? Beyond being academically interesting, does it matter that chaos can be observed in models such as the swinging of a jointed pendulum, or rising smoke? Those models demonstrate that chaos and complexity are part of nature, and bad only in context, as in a typhoon, or when wilfull, as in deliberately inflicting damage and pain in war. What if increased awareness of such phenomena constrained commanders in action, presenting an opportunity for foes knowing nothing of such elegant concepts, whose urge to act was not impaired by a sense of ambiguity, or who comprehended them as an underlying logic and bent that to their use. If careful analysis of such phenomena yielded advantages, and both adversaries employed such stratagems, that might, like the mutual breaking of codes in the Battle of the Atlantic in World War II, neutralize each others' ploys, or generate greater complexity—or chaos.

Considering ramifications of chaos-complexity leads to whorls of ambiguity far different than the orderly lines, grids, and exact measurements esteemed in the sciences and policy sciences. As noted earlier, in the West, visions of orderly trajectories, lines, and cycles[50] and of full, coherent consciousness are deeply seated in language and style, along with the assertive posturing that has often counted for much in war and military history. That is reflected in popular expressions like "right down the line," "straight as a die," "straightforward," and "stand-up kind of guy." The indeterminacy and chaos of war have often been offset by the confident demeanor of commanders, or by chroniclers afterward, regardless of whether their knowledge was proportional. Chaos and complexity theories do offer some potential to sort out bluff and bluster, and might allow some clearer consideration, both in praxis and in analysis, of what correlations are between factors, beyond loose conclusions or claims of exact perception and control of intricate processes.

Since historians, like commanders and staffs, rarely see more than a flickering, skeletal facsimile of the elaborate processes of warfare, military history resembles the gathering of interviews from survivors of a disaster added to journalistic accounts and observers' and rescue workers' testimony to comprise a report. Such synthesis may provide a broader and more detailed view of what happened than any individual account, but it will also fall short of absolute truth, and to what degree it does is also immeasurable. That paradox, depicted in the Japanese tale and film *Rashomon* and in Robie Macaulay's novel *The Disguises of Love,* analogically moves military history, and especially portrayals of warfare, closer to impressionist painting than to realism.

Viewing such dilemmas in the light of complexity and chaos leads to the question of how much absolute limits impede the tracing of war dynamics and the framing of laws and principles. Lack of surety has not prevented the waging of war, the preparing for it, the formulation of policy and doctrine, or the striking of postures of certainty. Is warfare then an aggregate problem that can never be solved, even by improved concepts and techniques? Since chaos-complexity theories, like "fuzzy sets," have been used to bring what once appeared to be random turbulence and anomalies into the realm of rationalistic description and in some cases utility, the main concern here is to determine whether the sense of war-as-chaos arises from its really being so, absolutely and irreducibly, or whether that only reflects limits of methods of sensing or inadequate conceptualization.[51]

Just as the surge of interest in nonlinear phenomenology and research in the 1970s stemmed from an increased capacity of computers to process or "crunch" experimental data, the ongoing evolution of computer technology and refined methodologies may further reduce ambiguities and allow dimensions of warfare previously out of view or wholly unrecognized to be identified, controlled, and understood in hindsight and prediction. That is very difficult to foresee, since, previously, the development of new scientific tools such as microscopes, telescopes, and particle accelerators produced both a clearer understanding of some fundamental processes, and a heightened awareness of further orders of complexity and uncertainty. That has also been seen in the case of military command-and-control, from the appearance of highly accurate printed maps in the mid-eighteenth century through the coming of railways and telegraphy to the air defense and strategic command-and-control systems of the late twentieth century. In stages, the rationalizing of data arrays permeated all command processes, from small-unit tactics through logistics to the orchestration of literally global operations. Operational maps in headquarters became intricate arrays of annotations, boundaries, routes, arrows, and bloc symbols at the center of intricate radians of communication nets. Those, although far more detailed than maps in most military histories, did not convey the full complexity of the vast number of individual "particles"—people, ships, aircraft, vehicles, guns, tanks, and so forth—that compose the actual fabric of armed forces. The bloc symbols and notations on military situation maps in ground warfare did not indicate readily such quantities or qualities as units' strength, health, logistical levels, time in combat, casualty history, fatigue, morale indicators, and so forth. Such data was often presented in briefings, available in staff files, or floating about in commanders' or staff officers' memories.

In the last quarter of the twentieth century, advances in computer display technology recast the graphic portrayal of battle dynamics, increasing the degree of detailing, and altered military professionals' imaging of war. That awareness was visible in television reports during Operation DESERT STORM, but the full effect of that is not fully apparent.[52] The new graphics added yet another level of complexity in the form of increased information flow that further exceeded humans' capacity to monitor it.[53] Increased resolving power of radar and computers and satellite-linked communications allowed detailed perception and identification of battle elements, including great swarms of individual vehicles on both sides, but did not prevent "friendly fire" incidents, or misidentification. The postmortems of the invasion of Panama in late 1989 and of the Gulf War also made it clear that planning and orchestration had fallen well short of perfection, even against a foe lacking a major electronic warfare capacity.[54]

While detailed and rapid imaging of combat offered obvious advantages, those wars reflected the persistence of a gap between the ability to monitor and actually control operations. Since the mid-nineteenth century, each stage of technical refinement yielded an increase in complexity and chaos in warfare and stimulated a search for antidotes and correctives. By the mid-1980s, filtering, buffering, sorting, keying, and denoting software had come into use in command-and-control systems. Leaps in computational speed and detail that allowed simulations and models to approach operational or "virtual" reality cast in doubt Theodore von Karman's assertion of the late 1950s that "no weapon or weapons system can be tested in theory . . . [but] only through an analysis of actual operations,"[55] although such technical advances did not end skepticism about sophisticated electronic systems' effectiveness, and left open the debate among military professionals about the relative importance of fighting spirit versus technology.[56] Just as U.S. Army junior officers in the Vietnam War resented high-level commanders in helicopters using radios to override subordinates' authority—the so-called "squad leader-in-the-sky"—after the Gulf War, military and intelligence professionals and members of the "Reform Movement" also critiqued "C-cubed" technology.[57] That ongoing tension and the normative resistance to novelty in military circles may retard the drawing of implications from chaos-complexity, but a debate has already begun, with some analysts seeing such concepts as relevant to defense-related problems,[58] and others rejecting their applicability.[59]

Just as considering the relevance of chaos-complexity theoretics in the sciences has generated expanding webs of ramifications, reexamining military history from nonlinear perspectives may generate intellectual currents and models that alter academic conceptions of war, or those of military professionals as they draw upon military history for models. It is not possible to foresee which strands of the conceptual net would bear the most traffic. While applying such theories might augment tactical effectiveness, a heightened awareness of war's complexity might also lead to a realization of terminal chaos. Following implications to that most darkling plain, the ongoing "evolution" and diversification of warfare portends an unknown and perhaps unknowable "omega." Puzzling out a remedy or designing a brake to that momentum obviously requires much rigor and will. Especially perplexing is the possibility that weakly perceived and controlled fragmentary factors such as arms transfers, regional frictions, espionage, and propaganda may all be subtle charges to a kind of historic capacitor that may, with little or no warning, from a state of apparent stability suddenly discharge with great and disproportionate violence, as the Belle Epoque exploded into the Great War. Not as much concern is being focused on that dilemma as was the case when superpower polarity placed the world on the edge of a thermonuclear precipice. Looking past that grim contingency, chaos-complexity theoretics offers some hope in underscoring how individual will and action might have far more effect than the framers of or believers in broad, linear general laws and processes would have believed—or would have been willing to believe.

NOTES

1. E.g., Richard P. Hallion, *Storm Over Iraq: Air Power and the Gulf War* (Washington, D.C.: Smithsonian Institute Press, 1992), pp. 2–3, and Eliot Cohen, "A GWAPS Primer," in Eliot Cohen, et al., eds., *Highlights: Gulf War Air Power Survey* (Washington, D.C.: U.S. Air Force, 1993), p. 13.

2. For a thoughtful essay on the problem of failure, see Walter Laqueur, *A World of Secrets: The Uses and Limits of Intelligence* (New York: Basic Books, 1985), pp. 255–292; for a recent overview of relevant literature, see Thomas Powers, "The Truth About the CIA," *New York Review of Books*, May 13, 1993, pp. 49–56.

3. E.g., most of implications drawn from the Korean War by the U.S. Air Force did not apply in the 2nd Indochina War, cf. the conclusion to Frank Futrell's *The U.S. Air Force in the Korean War* (Washington, D.C.: Office of Air Force History, 1961).

4. E.g., the altering of German military documents during the Weimar period, the extended sequestration of records in various archives, routinized file "purges" in bureaucracies, the extensive destruction of intelligence-related materials in Japan during the U.S. occupation, the Howard Hughes autobiography and Adolf Hitler diary forgeries, and the energetic "shredding" in the Iran-Contra affair.

5. Margaret J. Wheatley, *Leadership in the New Science: Learning About Organizations for an Orderly Universe* (San Francisco: Berret & Koehler, 1992), p. 7.

6. Quoted in Peter Vansittart, ed., *Voices From the Great War* (New York: Discus Books, 1985), p. 81.

7. Burt Kosko, *Fuzzy Thinking: The New Science of Fuzzy Logic* (New York: Hyperion Books, 1993), p. 8.

8. John Lewis Gaddis, *The United States and the Origins of the Cold War, 1941–1947* (New York: Columbia University Press, 1972), p. 360.

9. E.g., Occam's razor; also see Frank A. Gagliardi and Cheryl A. Doughan, "Predictive Technology: Method or Madness?" *National Defense*, 77/487 (April 1993), p. 14.

10. While simulations are widely seen as a major boon to training and a substitute for readiness, they also constitute a variant of doctrinal fixation, to the extent that "gaming and simulation can . . . reinforce biases and narrow the span of an organization's attention," cf. Paul Bracken, *Gaming in Hierarchical Defense Organizations* (New Haven: Yale School of Organization and Management, 1987), p. 2.

11. E.g., see Stanley D. Rosenberg, "The Threshold of Thrill: Life Stories in the Skies Over Southeast Asia," in *Gendering War Talk*, ed. Miriam Cooks & Angela Woollacot (Princeton: Princeton University Press, 1993), pp. 44–46 and 63–64.

12. E.g., see Fredric Smoler, "The Secret of the Soldiers Who Didn't Shoot," *American Heritage*, 40:2 (March 1989), pp. 36–45.

13. An infantry combat veteran of World War II caught the essence of that dilemma in recounting, "Lots of times you don't know. You're out there in the field and everybody's firing, eyeball to eyeball, and you don't know who got hit by who," cf. Richard M. Stannard, *Infantry: An Oral History of a World War II American Infantry Battalion* (New York: Twayne Publishers, 1993), p. 57.

14. For a lengthy and very useful exploration of the dimensions of this set of problems, see John Keegan, *The Face of Battle* (New York: Viking Press, 1976), pp. 15–78.

15. An exception is John Costello, *Virtue Under Fire: How World War II Changed Our Social and Sexual Attitudes* (New York: Fromm International, 1987).

16. E.g., see Sanitsuda Ekachai, "Woman and War," *World Press Review*, February 1993, p. 47.

17. E.g., see J.P. Wood, *Aircraft Nose Art: 80 Years of Aviation Artwork* (New York: Crescent Books, 1982); for a journalistic essay on the question, see William Broyles, Jr., "Why Men Love War," *Esquire*, 102:5 (November 1984), pp. 55–58, 61–62, and 65.

18. E.g., see Lewis Yablonsky's description of the "we" feeling in *The Violent Gang* (Baltimore: Penguin Books, 1970), p. 188.

19. The blend of prison and military cultures in "boot camp" correctional programs has been widely viewed as a natural fusion.

20. Relatively well known exceptions are Jaroslav Hasek's *The Good Soldier Schweik*, and Sergeant Bilko.

21. E.g., *Grand Illusion, From Here to Eternity, Mister Roberts, Bridge on the River Kwai, To Hell and Back, Tunes of Glory,* and *An Officer and a Gentleman.*

22. E.g., Hitler, Mussolini, Batista, and Idi Amin were former NCOs; Peron, Naguib, and Nasser, colonels; and Muammar Qaddafi, a captain.

23. A model exploration of this underconsidered area is Dennis E. Showalter's "Caste, Skill, and Training: The Evolution of Cohesion in European Armies from the Middle Ages to the Sixteenth Century," *Journal of Military History*, 75:3 (July 1993), pp. 407–430.

24. E.g., Karl von Clausewitz, Friedrich Engels, Billy Mitchell, Felix Steiner, Tom Wintringham, Evans Carlson, Mikhail Tukhachevski, and James Channon.

25. Woodrow Wilson wrote a brief essay assailing that practice, and Franklin Roosevelt was the only head of a major combatant nation in the Second World War who did not wear a uniform, although he often wore a naval officer's boat-cloak.

26. For a recent survey, see Jerrold M. Post and Robert S. Robins, *When Illness Strikes the Leader: The Dilemma of the Captive King* (New Haven: Yale University Press, 1993).

27. For an attempt to bestride the rational-irrational chasm, see Thomas C. Schelling, *The Strategy of Conflict* (Cambridge, Mass.: Harvard University Press, 1980).

28. E.g., see Doris Kearns, *Lyndon Johnson and the American Dream* (New York: Signet Books, 1977), pp. 419–425.

29. E.g., V. V. Druzhinin & D.S. Kontorov, *Concept, Algorithm, Decision* (Washington, D.C.: U.S. Government Printing Office, n.d.) [c. 1978], esp. pp. 179–199; for an example of theoretical work across a range of scientific disciplines, see A. N. Tikhnonov & A.V. Goncharsky, *Ill-Posed Problems in the Natural Sciences* (Moscow: MIR Publishers, 1987); for a lexicon of concepts, see Nathan Leites, *Soviet Style in War* (New York: Crane & Russak, 1982).

30. For basic Soviet definitions, see N.V. Shislina, ed., *Sovremennaya Ideologicheskaya Borba* (Moscow: Izdatelstvo Politicheskoi Literatur, 1988), pp. 83–85, and M.I. Kondakova & A.S. Vishnyakova, eds., *Kratkii Pedagogicheskii Slovar*

Propagandista (Moscow: Izdatelstvo Politicheskoi Literatur, 1988), pp. 49–50; T. Vlasova, *Marxist-Leninist Philosophy* (Moscow: Progress Publishers, 1987), pp. 45–75; and V. Krapavin, *What is Dialectical Materialism?* (Moscow: Progress Publishers, 1985).

31. E.g., see A.A. Voronov, "The Decomposition Approach to Multiconnected Systems," *Management and Control in Large Systems* (Moscow: MIR Publishers, 1986), pp. 50–71; for a recent biographical perspective, including some links between control authority and the organs of state security, see E.B. Koritskii, Y.A. Lavrikov & A.M. Omarov, *Sovyetskaya Upravlencheskaya Mysl, 20-X Godov: Kratkii Immenoi Spravochnik* (Moscow: "Ekonomika," 1990).

32. "Dialectical materialism," in Tom Bottomore, ed., *A Dictionary of Marxist Thought* (Cambridge, Mass.: Harvard University Press, 1983), p. 120.

33. Alan Beyerchen, "Clausewitz, Non-Linearity, and the Unpredictability of War," *International Security*, 17:3 (Winter 1992/93), pp. 59–90.

34. See Martin Van Creveld, "Trinitarian War," pp. 49–57 in *The Transformation of War* (New York: Free Press, 1991).

35. Stephen H. Kellert, *In the Wake of Chaos: Unpredictable Order in Dynamical Systems* (Chicago: University of Chicago Press, 1993), p. 33.

36. Deemed a conceptual "Dreadnought" by Roy Soncrant in his seminal and insightful unpublished paper, "The Implications of Complexity."

37. E.g., see Adam Szladow & Wojciech Ziark, "Rough Sets: Working With Imperfect Data," *A.I. Expert*, 8:7 (July 1993), pp. 36–41, and Burt Kosko & Saturo Isaka, "Fuzzy Logic," *Scientific American*, 269:1 (July 1993), pp. 76–81.

38. E.g., see Soncrant, "Implications of Complexity," p. 23.

39. Prime examples of the latter are Ward Moore's *Bring the Jubilee*; Philip K. Dick's *The Man in the High Castle*; Sarban's *The Sound of His Horn*; Keith Roberts' *Pavane*; and Robert Silverberg's *Up the Line*.

40. Donald N. McCloskey, "History, Differential Equations, and the Problem of Narration," *History and Theory*, 21:1 (Winter 1991), pp. 21–36.

41. John Lewis Gaddis, "International Relations Theory and the End of the Cold War," *International Security*, 17:3 (Winter 1992/93), p. 52.

42. Peter R. Killeen, "Behavior as a Trajectory Through a Field of Attractors," in J.R. Brink & C.R. Haden, eds., *The Computer and the Brain: Perspectives on Human and Artificial Intelligence* (North Holland: Elsevier Science Publishers, 1989), p. 54.

43. Trevor N. Dupuy, Curt Johnson, David L. Bongard & Arnold C. Dupuy, *How to Defeat Saddam Hussein: Scenarios and Strategies for the Gulf War* (New York: Warner Books, 1991).

44. John N. Warfield, *An Assault on Complexity* (Seattle: Battelle Research Institute, 1973), pp. ii–iii.

45. Although arguing that cause and effect can be determined in detailed analysis of battles, U.S. Army colonel Harold Nelson pointed out that "wills in opposition" and "fog and friction" create a "dynamic, interactive environment," cf. Harold Nelson, "What the Staff Ride Can Depict: Face of Battle, Clash of Wills and Arms, Generalship, and Cause and Effect," *The Army Historian*, October 1988 [PB-20-88-1 (No 12)], p. 15.

46. Some complexity-chaos enthusiasts have, like some General Systems Theory, cybernetics, and quantum physics enthusiasts, been drawn toward the

metaphysical and mystical, e.g., see Frank A. Gagliardi & Cheryl A. Doughan, "Predictive Technology: Method or Madness?" *Discover*, June 1991.

47. E.g., see Clifford Pickover, *Computers and the Imagination* (New York: St. Martin's, 1991), pp. 307–08 and 311–312.

48. See Burt Kosko and Saturo Isaka, "Fuzzy Logic," *Scientific American*, 269:1 (July 1993), pp. 76–81.

49. E.g., see Ralph Strauch, *The Reality Illusion: How We Create the World We Experience* (Wheaton, Ill.: Quest Books, 1983).

50. For an incisive critique of war cycle analyses, see Brian J.L. Berry, *Long-Wave Rhythms in Economic Development and Political Behavior* (Baltimore: Johns Hopkins University Press, 1991), pp. 156–166.

51. Stephen H. Kellert, *In the Wake of Chaos: Unpredictable Order in Dynamical Systems* (Chicago: University of Chicago Press, 1993), p. 44.

52. E.g., see Al Campen, ed., *The First Information War* (Fairfax, Va.: Armed Forces Communication and Electronics Association Press, 1993).

53. There has been little public discussion of the fact that the blinding speed of the Strategic Defense Initiative—"Star Wars"—space-based missile defense systems' computers precluded human controllers being involved in battle management.

54. E.g., see "Tacitus," "Few Lessons Were Learned in Panama Invasion," *Armed Forces Journal International*, 130:11 (June 1993), pp. 54–56, and James G. Burton, "Pushing Them Out the Back Door," *U.S. Naval Institute Proceedings*, 119:6 (June 1993), pp. 37–42.

55. Theodore von Karman, et al., *Operational Research in Practice: Report of a NATO Conference* (London: Pergamon Press, 1958), p. 8.

56. E.g., see Crosbie E. Saint, "Thought Stimulation Versus 'Stillborn' Solutions," *Army*, 43:6 (June 1993), pp. 69–70.

57. For a brief against the structural blending of command-and-control and intelligence, see Walter Jajko, *The Future of Defense Intelligence* (Washington, D.C.: Consortium for the Study of Intelligence, 1993), pp. 11–13.

58. Alvin Saperstein, "SDI: A Model for Chaos," *Bulletin of the Atomic Scientists*, 44:8 (October 1988), pp. 40–43.

59. Jack Snyder & Robert Jervis, *Coping With Complexity in the International System* (Boulder: Westview, 1993), p. 12.

Select Bibliography

BOOKS

Alexander, John B., Richard Groller and Janet Morris. *The Warrior Edge.* New York: William Morrow, 1990.

Alger, John I. *The Quest for Victory: The History of the Principles of War.* Westport, Conn.: Greenwood Press, 1982.

Angelucci, Enzo, ed. *The Rand McNally Encyclopedia of Military Aircraft, 1914–1980.* New York: The Military Press, 1983.

Arquilla, John, and David Ronfeldt. *Cyberwar Is Coming!* Santa Monica, Calif.: RAND Corporation, 1992.

Ashby, W. Ross. *Design for a Brain: The Origin of Adaptive Behavior.* London: Chapman & Hall, 1960.

Atteridge, Andrews Hilliard. *The German Army in War.* London: Methuen, 1915.

Bassford, Christopher. *Clausewitz in English: The Reception of Clausewitz in Britain and America.* Oxford: Oxford University Press, 1994.

Beer, Stafford. *Decision and Control.* London: John Wiley & Sons, 1966, pp. 38–39.

Berlin, Isaiah. *Four Essays on Liberty.* London: Oxford University Press, 1954.

———. *Historical Inevitability.* London: Oxford University Press, 1955.

Berman, Larry. *Planning a Tragedy: The Americanization of the War in Vietnam.* New York: W.W. Norton, 1982.

Berry, Brian J.L. *Long-Wave Rhythms in Economic Development and Political Behavior.* Baltimore: Johns Hopkins University Press, 1991.

Bohm, David, and F. David Peat. *Science, Order and Creativity.* Toronto: Bantam Books, 1987.

Boylan, Bernard. *Development of the Long-Range Fighter Escort* (U.S. Air Force Historical Studies No. 136). Maxwell Air Force Base: Research Studies Institute, U.S. Air Force Historical Division, 1955.

Briggs, John, and F. David Peat. *Turbulent Mirror: An Illustrated Guide to Chaos Theory and the Science of Wholeness.* New York: Harper & Row, 1989.

Buell, Thomas. *Master of Sea Power.* Boston: Little Brown, 1980.

Burckhardt, Jacob. *Reflections on History.* Indianapolis: Liberty Classic, 1975.

Campbell, N.J.M. *Jutland: An Analysis of the Fighting.* Annapolis: U.S. Naval Institute Press, 1986.

Charters, David A., Marc Milner, and Brent Wilson, eds. *Military History and the Military Profession.* Westport, Conn.: Praeger, 1992.

Coffey, Thomas M. *Hap: Military Aviator.* New York: Viking Press, 1982.

Collingwood, R. G. *Essay on Metaphysics.* Chicago: Henry Regnery, 1972.

Commager, Henry Steele, ed. *The Blue and the Gray.* Indianapolis: Bobbs-Merrill, 1950.

Condensed Analysis of the Ninth Air Force in the European Theater of Operations. Washington, D.C.: Office of Air Force History, 1984 (orig. publ. 1946), p. 48.

Copp, DeWitt S. *A Few Great Captains.* Garden City, N.Y.: Doubleday, 1980.

Cru, Jean Norton. *War Books: A Study in Historical Criticism.* Edited and translated by Stanley J. Pincentl. San Diego: San Diego State University Press, 1988.

Davies, D.R., and G.S. Tune. *Human Vigilance Performance.* New York: American Elsevier, 1969.

de Bono, Edward. *Practical Thinking.* London: Jonathan Cape, 1971.

de Seversky, Alexander P. *Air Power: Key to Survival.* New York: Simon & Schuster, 1950.

Doughty, Robert. *The Evolution of U.S. Army Tactical Doctrine, 1946–1976.* Fort Leavenworth: Combat Studies Institute, 1979.

Douglass, Joseph D., Jr., and Amoretta Hoeber, eds. *Selected Readings from "Military Thought."* Washington, D.C.: U.S. Government Printing Office, c. 1980 (U.S. Air Force Studies in Communist Affairs, Vol. 5, Pt 2).

Douhet, Giulio. *The Command of the Air.* Trans. by Dino Ferrari. New York: Coward McCann, 1942.

Dray, William. *Laws and Explanations in History.* Oxford: Oxford University Press, 1957.

Druzhinin, V.V., and D.S. Kontorov, *Decision Making and Automation: Concept, Algorithm, Decision.* Washington, D.C.: U.S. Government Printing Office, n.d. (c. 1978).

Eckert, Edward K., ed. *In War and Peace: An American Military History Anthology.* Belmont, Calif.: Wadsworth Publishing Co., 1990.

Eigen, Manfred and Ruthild Winkler. *Laws of the Game: How the Principles of Nature Govern Chance.* New York: Harper Colophon, 1981.

Ellis, John. *The Sharp End: The Fighting Man in World War II.* New York: Charles Scribner's, 1980.

Epstein, Robert M. *Napoleon's Last Victory and the Emergence of Modern War.* Lawrence: University of Kansas Press, 1993.

Evans, Joseph, ed. *On the Philosophy of History.* New York: Charles Scribner's Sons, 1957.

FM 100-5 Operations. Washington, D.C.: Department of the Army, 1993.

Fabyanic, Thomas A. *Strategic Air Attack in the United States Air Forces: A Case Study* (Air War College/Air University Report No. 5899). Manhattan, Kans.: Military Affairs/Aerospace Historian, 1976.

Feder, Jens. *Fractals.* New York: Plenum Press, 1988.

Ferguson, Yale H. and Richard H. Mansbach. *The State, Conceptual Chaos, and the Future of International Relations.* Boulder, Colo.: L. Rienne, 1989.
Francillon, Rene. *Japanese Aircraft of the Pacific War.* London: Putnam, 1970.
Freiden, Seymour and William Richardson, eds. *The Fatal Decisions.* New York: William Sloane Associates, 1956.
Friedman, Norman. *U.S. Battleships: An Illustrated Design History.* Annapolis, Md.: U.S. Naval Institute Press, 1985.
Frightfulness in Retreat. London: Hodder & Stoughton, 1917.
Fuller, J.F.C. *Machine Warfare: An Inquiry into the Influence of Mechanics on the Art of War.* Washington, D.C.: Infantry Journal Press, 1943.
Gardner, Howard. *Creating Minds: An Anatomy of Creativity Seen Through the Lives of Freud, Einstein, Picasso, Stravinsky, Eliot, Graham and Gandhi.* New York: Basic Books, 1993.
Gat, Azar. *The Origins of Military Thought from the Enlightenment to Clausewitz.* Oxford: Clarendon, 1989.
Gleick, James. *Chaos: Making a New Science.* New York: Penguin Books, 1987.
Goerlitz, Walter. *Paulus and Stalingrad.* New York: Citadel Press, 1963.
Gregory, R. L. *The Intelligent Eye.* New York: McGraw-Hill, 1970.
Gupta, Madan M. and Ellie Sanchez. *Fuzzy Information and Decision Processes.* Amsterdam: North Holland Publishing Co., 1982.
Hagan, Kenneth. *This People's Navy: The Making of American Sea Power.* New York: Free Press, 1991.
Halberstam, David. *The Best and the Brightest.* New York: Ballantine Books, 1992.
Halley, James J. *The Role of the Fighter in Air Warfare.* Edited by Charles W. Cain. Garden City: Doubleday & Co., 1971.
Hawking, Stephen. *A Brief History of Time: From the Big Bang to Black Holes.* Toronto: Bantam Books, 1988.
Hayles, N. Katherine. *Chaos Bound: Orderly Disorder in Contemporary Literature and Society.* New York: Cornell University Press, 1990.
Helmbold, Robert L. *Decision in Battle: Breakpoint Hypotheses and Engagement Termination Data.* Santa Monica: Rand Corporation, 1971.
Henderson, G.F.R. *The Science of War: A Collection of Essays and Lectures, 1891–1903.* London: Longmans Green, 1912.
Hill, R.J.T. *Phantom Was There.* London: E. Arnold, 1951.
Holmes, Richard. *Acts of War: The Behavior of Men in Battle.* New York: Free Press, 1985.
Horricks, Raymond. *Military Mindlessness: An Informal Compendium:* New Brunswick: Transaction Publishers, 1993.
Jomini, Antoine. *The Art of War.* Translated by G.H. Mendell and W.P. Craighill. Westport, Conn.: Greenwood Press, 1971; orig. publ. 1862.
Kahn, Herman. *On Escalation.* New York: Frederick Praeger, 1965.
Keegan, John. *The Face of Battle.* New York: Viking Press, 1976.
Kent, Sherman. *Strategic Intelligence for American World Policy.* Hamden, Conn.: Archon Books, 1965.
Kersh, Gerald. *Sergeant Nelson of the Guards.* Philadelphia: John C. Winston, 1945.
Lanchester, F.W. *Aircraft in Warfare: The Dawn of the Fourth Arm.* New York: Appleton, 1916.
Lanier-Graham, Susan D. *The Ecology of War.* New York: Walker & Co., 1993.

Lawrence, T.E. *Seven Pillars of Wisdom: A Triumph.* Garden City, N.Y.: Doubleday Doran, 1937.

Lazarus, Richard. *Psychological Stress and the Coping Process.* New York: McGraw-Hill, 1966.

Lewin, Roger. *Life at the Edge of Chaos.* New York: Macmillan, 1992.

Liddell Hart, B.H. *The Real War 1914–1918.* Boston: Little Brown, 1930.

———. *Strategy.* New York: Praeger, 1965.

———, ed. *The Rommel Papers.* New York: Harcourt, Brace & Co., 1953.

Livingston, William L. *The New Plague: Organizations in Complexity.* Bayside, N.Y.: F.E.S. Limited Publishers, 1985.

Lucas-Phillips, C. E. *Alamein.* Boston: Little Brown, 1962.

Mackay, Ruddock F. *Fisher of Kilverstone.* Oxford: Clarendon, 1937.

McKnight, Frank H. *Risk, Uncertainty and Profit.* Boston: Houghton Mifflin, 1921.

Mansfield, Sue. *The Gestalts of War.* New York: Dial Press, 1982.

Marcus, John J. *Heaven, Hell and History.* New York: Macmillan, 1967.

March, James G., Roger Weissinger, Baylon Ryan, and Pauline Ryan. *Ambiguity and Command: Original Perspectives on Military Decision-Making.* Marshfield, Mass.: Pitman Publishing Co., 1986.

Marder, Arthur J. *The Anatomy of British Sea Power: A History of British Naval Policy in the Dreadnought Era, 1880–1905.* Hamden: Archon, 1964.

———. *From the Dreadnought to Scapa Flow: The Royal Navy in the Fisher Era, 1904–1919,* Vol. 3, *Jutland and After (May 1916–December 1916).* London: Oxford University Press, 1960.

Nicolis, Gregoire and Ilya Prigogine, *Ordering Complexity: An Introduction.* New York: Freeman & Co., 1989.

Ogorkiewicz, Richard M. *Armoured Forces: A History of Armoured Forces and Their Vehicles.* New York: Arco Publishing Co., 1970.

Pagels, Heinz R. *The Dreams of Reason: The Computer and the Rise of Science of Complexity.* New York: Simon & Schuster, 1988.

Parton, James. *"Air Force Spoken Here": General Ira Eaker and the Command of the Air.* Bethesda, Md.: Adler & Adler, 1986.

Paxson, E.W., M.G. Weiner and R.A. Wise, *Interactions Between Tactics and Technology in Ground Warfare.* Santa Monica: Rand Corporation, 1979 (R–2377–ARPA).

Peat, F. David. *The Philosopher's Stone: Chaos, Synchronicity and the Hidden Order of the World.* New York: Bantam Books, 1991.

Pick, Daniel. *The War Machine: The Rationalisation of Slaughter in the Modern Age.* New Haven, Conn.: Yale University Press, 1993.

Poundstone, William. *The Renaissance Universe: Cosmic Complexity and the Limits of Scientific Knowledge.* Chicago: Contemporary Books, 1985.

Prigogine, Ilya. *From Being to Becoming: Time and Complexity.* New York: W.H. Freeman, 1980.

———and Isabelle Stengers. *Order Out of Chaos: Man's New Dialogue With Nature.* Toronto: Bantam Books, 1988.

Quade, E.S. *Analysis for Military Decisions.* Santa Monica: Rand Corporation, 1964 (R–387–PR).

Richardson, F.M. *Fighting Spirit: A Study of Psychological Factors in War.* London: Leo Cooper, 1978.

Roeder, George H., Jr. *The Censored War: American Visual Experience During World War II.* New Haven, Conn.: Yale University Press, 1993.

Rommel, Erwin. *Attacks.* Vienna,Va.: Athena Press, 1979.

Rosenau, James. *Turbulence in World Politics.* Princeton: Princeton University Press, 1990.

Ruelle, David. *Chance and Chaos.* Princeton: Princeton University Press, 1991.

Russett, Cynthia Eagle. *The Concept of Equilibrium in American Social Thought.* New Haven, Conn.: Yale University Press, 1966.

Schroeder, Manfred. *Fractals, Chaos, Power Laws.* New York: W. H. Freeman, 1991.

Smith, Dale O. *U.S. Military Doctrine: A Study and Appraisal.* New York: Duell, Sloan & Pearce, 1955.

Smith, Peter. *British Battlecruisers.* New Malden, Surrey: Almanack Press, 1962.

Stewart, Ian. *Does God Play Dice? The Mathematics of Chaos.* Cambridge: Blackwell, 1991.

Stockfish, J.A. *Models, Data and War: A Critique of the Study of Conventional Forces.* Santa Monica: Rand Corporation, 1975 (R–1526–PR).

Strauch, Ralph. *The Reality Illusion: How We Create the World We Experience.* Wheaton, Ill.: Quest Editions, 1983.

Taylor, Calvin W. and Frank Barron. *Scientific Creativity: Its Recognition and Development.* New York: John Wiley & Sons, 1963.

Todd, William. *History as Applied Science.* Detroit: Wayne State University, 1972.

Traub, J. F., G. W. Wasilowski and H. Wozniakowski. *Information, Uncertainty, Complexity.* London: Addison-Wesley, 1983.

Vagts, Alfred. *A History of Militarism.* New York: Free Press, 1959.

von Moltke, Helmuth Karl Bernhard. *Gedanken von Moltke.* Berlin: Atlantik Verlag, 1941.

Waldrop, M. Mitchell. *Complexity: The Emerging Science at the Edge of Order and Chaos.* New York: Simon & Schuster, 1992.

Weigley, Russell F. *The American Way of War.* New York, Macmillan, 1973.

———. *Eisenhower's Lieutenants: The Campaign of France and Germany, 1944–1945,* Vol. 1. Bloomington: University of Indiana Press, 1981.

Wilkinson, David. *Deadly Quarrels: Lewis F. Richardson and the Statistical Study of War.* Berkeley: University of California Press, 1980.

Wolff, Leon. *In Flanders Fields.* New York: Viking Press, 1958.

Wolk, Herman S. *Planning and Organizing the Postwar Air Force.* Washington, D.C.: Office of Air Force History, 1984.

Wright, Quincy. *A Study of War.* Chicago: University of Chicago Press, 1967.

ARTICLES

Aref, Hassan and Gretar Tryggvason, "Vortex Dynamics of Passive and Active Interfaces." Paper prepared for Conference on Fronts, Interfaces and Patterns, May 2–6, 1983, Center for Non-Linear Studies, 29 pp.

Arthur, W. Brian. "Why Do Things Become More Complex?" *Scientific American* 268:5 (May 1993): 144.

Artigiani, Robert. "Post-Modernism and Social Evolution: An Inquiry." *World Futures* 30:3 (Autumn 1991): 149–161.

Bak, Per and Kan Chen. "Self-Organized Criticality." *Scientific American* 264:1 (January 1991): 46–53.

Batterman, Robert. "Randomness and Probability in Dynamic Theories: On the Proposals of the Prigogine School." *Philosophy of Science* 53:2 (June 1991): 241–261.

Boylan, Bernard. "The Search for a Long-Range Escort Plane, 1919–1941." *Military Affairs* 30:2 (Summer 1966): 58–59.

Chennault, Claire. "The Role of Defensive Pursuit," in *Coast Artillery Journal*, Pt. 1, 76:6 (November–December 1933): 412–416; Pt. 2, 77:1 (January–February 1934): 5–11; and Pt. 3, 77:3 (March–April 1933): 87–93.

Clarke, Judith. "First Punches: Review of Analogies at War: Korea, Munich, Dien Bien Phu and the Vietnam Decisions of 1965." *Far Eastern Economic Review* (December 3, 1992): 3.

Cole, Hugh. "War's Forgotten Men." *Vignettes* No. 69, U.S. Army Center for Military History (April 18, 1977).

Corcoran, Elizabeth. "The Edge of Chaos." *Scientific American* 267:4 (October 1992): 17, 18, 20, and 22.

———. "Ordering Chaos," *Scientific American* 265:2 (August 1991): 96 and 98.

Culham, Phyllis. "Chance, Command and Chaos in Ancient Military Engagements." *World Futures* 27:3 (Autumn 1989): 191–205.

Dewar, Alfred F. "The Reorganization of the Naval Staff, 1917–1919." *Naval Review* 9:9 (September 1921): 182–188.

Dromi, Uri. "The Risks of Doctrinal Stagnation." *IDF Journal* No. 14 (Spring 1988): 24.

Dunker, Kenneth F. and Basile G. Rabbat. "Why America's Bridges are Crumbling." *Scientific American* 269:3 (March 1993): 66–72.

DuPuy, Trevor N. "Perceptions of the Next War." *Armed Forces Journal International* 118:10 (May 1980): 40, 50, and 54.

Emerson, William R. "Operation POINTBLANK: A Tale of Bombers and Fighters," in Harry R. Borowski, ed., *The Harmon Memorial Lectures in Military History, 1959–1987*. Washington, D.C.: Office of Air Force History, 1988, pp. 451–452.

Evans, Jonathan S. B. "Psychological Pitfalls in Forecasting." *Futures* 14:4 (August 1982): 258–259.

"From Catastrophe to Crisis." *Economist* 315/7654 (May 12, 1990): 85–86.

Funk, Paul E. "Battle Space: A Commander's Tool on the Future Battlefield." *Military Review* 73:12 (December 1993): 36–47.

Gilmore, G. Bernard. "Randomness and the Search for PSI." *Journal of Parapsychology* 54:1 (December 1989): 339.

Gordon, Theodore J. and David Greenspan. "Chaos and Fractals: New Tools for Technological and Social Forecasting." *Technological Forecasting and Social Change* 34:1 (October 1988): 1–25.

Grossman, Siegfried and Gottfried Mayer-Kress. "Chaos in the International Arms Race." *Nature* 337:8 (February 23, 1989): 701–704.

Hawkins, Charles F. "Modelling the Breakpoint Phenomena." *Signal* 43:11 (July 1989): 37–41.

Holley, I.B. "Concepts, Doctrines and Principles: Are You Sure You Understand Those Terms?" *Air University Review* 35:4 (July–August 1984): 91.

Hendrickson, L. and J. Myers. "Some Sources and Potential Consequences of Errors in Medical Data Recording." *Methods of Information in Medicine* 12:1 (January 1973): 38–45.

Hogan, John. "Profile: Reluctant Revolutionary: Thomas Kuhn Unleashed 'Paradigm' on the World." *Scientific American* 264:5 (May 1991): 40.

Holden, Constance. "Creativity and the Troubled Mind." *Psychology Today* 3:4 (April 1967): 9–10.

Jeffer, Edward K. "Generalizing: The Mystique of High Command." *Army* 43:6 (June 1993): 43–46.

Jones, R.V. "The Natural Philosophy of Flying Saucers," in Edward U. Condon, ed., *Final Report of the Scientific Study of Unidentified Flying Objects Conducted by the University of Colorado Under Contract to the United States Air Force.* New York: Bantam Books, 1969, p. 922–33.

Joynt, Carey B. and Nicholas Rescher. "The Problem of Uniqueness in History." *History and Theory* 1:11 (1961): 150–162.

Jurgens, Hartmut, Heinz-Otto Peitgen and Dietmar Saupe. "The Language of Fractals." *Scientific American* 263:2 (August 1990): 60–67.

Khalil, Elias L. "Natural Complex vs. Natural System" *Journal of Social and Biological Structures* 13:1 (January 1991): 11–31.

McCloskey, Donald N. "History, Differential Equations, and the Problem of Narrative." *History and Theory* 31:1 (Winter 1991): 21–36.

McNeill, William H. "Mythistory, or Truth, Myth, History, and Historians." *American Historical Review* 91:1 (February 1986): 1–10.

McQuie, Robert. "Military History and Mathematical Analysis." *Military Review* 40:5 (May 1970): 8.

Mainieri, Ronnie. "Impressions on Cycling." *Center for Non-Linear Studies Newsletter* No. 76 (March 1992): 1–5.

Mandelbaum, Maurice. "Historical Explanation: The Problem of 'Covering Laws.'" *History and Theory* 1:3 (1960): 229–242.

Mann, Steven R. "Chaos Theory and Strategic Thought." *Parameters: Journal of the U.S. Army War College* 22:3 (Autumn 1992): 54–67.

Marti, James. "Chaos Might Be the New World Order." *Utne Reader* 7:6 (November–December 1991): 30 and 32.

"Money and Mayhem." *Economist* 315/7651 (April 21, 1990): 93–94.

Morgan, M. Granger. "Risk Analysis and Management." *Scientific American* 269:1 (July 1993): 32–41.

Naeye, Robert. "Physics Watch: Chaos Squared." *Discover* 15:3 (March 1994): 28.

Nagel, Ernst. "Determinism in History." *Philosophy and Phenomenological Research* 20 (1960): 291–317.

Nayak, P.R. and J.M. Kettringham. "The Fine Art of Managing Creativity." *New York Times*, November 2, 1986, p. F 2.

Newell, Alan C. "The Dynamics and Analysis of Patterns," in D. Stein, ed., *Lectures in the Science of Complexity: SFI Studies in the Science of Complexity.* New York: Addison-Wesley Longmans, 1989, pp. 107–169.

Newman, William. "It's Simply a Matter of Scale." *Center for Nonlinear Studies Newsletter* No. 68 (July 1991): 1–9.

O'Quinn, Brian. "Strategic Change: 'Logical Incrementalism.'" *Sloan Management Review* 20:1 (Spring 1988): 24.

Ornstein, D. S. "Ergodic Theory." *Science* 243: 4588 (January 13, 1989): 182–187.

Pearl, Judea and Michael Tarsi. "Structuring Causal Trees." *Journal of Complexity* 2:1 (1986): 60–77.

Pfeiffer, W. and Arthur H. Westing. "Land War." *Environment* 13:9 (November 1971): 2–13.

Pool, Robert. "Chaos Theory: How Big an Advance?" *Science* 245/4913 (July 7, 1989): 26–28.

——. "Is It Chaos or Just Noise?" *Science* 243:4889 (January 6, 1989): 25–27.

Reisch, George A. "Chaos, History and Narrative." *History and Theory* 21:1 (Winter 1991): 1–20.

Ruthen, Russell. "Trends in Non-Linear Dynamics: Adapting to Complexity." *Scientific American* 268:1 (January 1993): 140.

Scharf, Rainer and Bala Sundaram. "Suppression of Chaos in Quantum Dynamics." *Center for Nonlinear Studies Newsletter* No. 77 (April 1992): 1–12.

Spiller, Roger. "Shell Shock." *American Heritage* (May/June, 1990): 75–87.

Stewart, Ian. "Does Chaos Rule the Cosmos?" *Discover* 13:11 (November 1992): 56–58 and 61–63.

Stivers, R.E. "The Mystique of Command Presence." *U.S. Naval Institute Proceedings* 94:8 (August 1968): 27–33.

Taylor, James. "Recent Developments in the Lanchester Theory of Combat," in K.B. Haley, ed., *OR 78*. Amsterdam: North Holland Publishing Co., 1979, p. 773.

Teplov, Boris. "K Voprosy O Praktischeskom Myshlenie." *Uchenye Zapiski* 5 (1945): 149–214.

Traub, Joseph F. and Henryk Wozniakowski. "Breaking Intractability." *Scientific American* 270:1 (January 1994): 102–107.

Wallace, Anthony F.C. "Revitalization Movements." *American Anthropologist* 58:2 (April 1956): 264–281.

Webster, Donovan. "The Soldiers Moved On. The War Moved On. The Bombs Stayed." *Smithsonian* 24:11 (February 1994): 26–37.

Westing, Arthur H., ed. *Environmental Hazards of War*. London: Sage Publications, 1990.

Westing, Arthur H. and E.W. Pfeiffer. "The Cratering of Indochina." *Scientific American* 226:5 (May 1972): 21–29.

Index

accident, in invention and discovery, 129
Admiralty, British, 51
aerial warfare, of attrition, 6; chivalry in, 99
aerospace warfare, 117
airborne forces, 101–102
Air Corps boards, 58, 62
Air Corps Tactical School. *See* United States—Army Air Corps Tactical School
air power, adherence to mass tactics, 109; non-linear potential of, 173
Air Staff, U.S. Army Air Corps/Air Forces, 64
Alexander, Harold (Field Marshal Earl), command style of, 102
algorithm, of combat, 91
Allenby, Edmund Henry Hynman (Field Marshal Viscount), command style of, 102; pacific views of, 158
Allied leaders, World War II, rhetoric of, 161–162
"alternative futures," 116, 171
alternatives, in planning, 40
amateurs, success in war, 107, 121
ambiguity: in defense planning, 155, 174; in warfare, 19, 117
American Civil War, 79, 107; casualties in, 9, 47, 99; medical services in, 106; momentum of, 99; Napoleonic models in, 121; command processes in, 160
American Revolution (1775–1783), 47, 86; depredations in, 108, 139; classical models in, 171
amphibious warfare, 98; in World War II, 90, 102; Jomini on, 110 n. 9
analogies, of chaos-complexity, 170
Anderson, Orvil (Maj.-Gen.), quoted, on fighter escort gap, 64
Andrews, Frank (Lt.-Gen.), calls for escort fighters, 1936, 59
"angle of attack" theory, 32, 117
"*Angriff Ohne Befehl*," Guderian's concept of, 104
Antietam, Battle of (1862), 20
antitank tactics, World War II, 54–56
Antwerp, air raids on, 64
Arab Revolt, 1916–18, 7. *See also* Lawrence, T.E
archaeology, of battle sites, 20
archival categorization, and distortion of history, 20, 105–106
Ardennes campaign (1944–45). *See* Bulge, Battle of the
arditi, Italian elite forces, World War I, 53
Armee de l'Air. *See* French Air Force
armored vehicles, light, 53–54

arms transfers, as complexity increment, 178
Army Air Corps. *See* United States—Army Air Corps
Army of Northern Virginia, in Wilderness campaign, 124
Army of the Potomac, in Wilderness campaign, 124
Arnold, Henry H. (Gen. of the Army): and fighter escorts, 43, 57, 58, 52, 59, 63
Arthur, W. Brian, on complexity as process, 107
art of war, and creativity, 115, 118, 126
Ashby, Ross, and requisite variety, 26, 50, 92
assassination, of leaders in World War II, 80
assault guns, 55
Assyrians, 138
Athens, Attican forests denuded by, 138
Atlantic, Battle of the (1939–1945); German Air Force-Navy cooperation, 153; code-breaking in, randomizing effect of, 175
atmospheric nuclear tests, 142
atrocities, in war, 99, 135, 139; Nazis', 161
attack aviation concept, abandoned, 68
Attica, forest denuded by Athens, 138
Attila the Hun, 116; environmental devastation by, 138
attractors, and doctrinal debates, 49. *See also* double attractors
auftragsbefehl/taktik (mission-type orders/tactics), 8, 79
Australian Corps, in France, 115; in August, 1918 offensive, 107; "peaceful penetration" raids of, 93
auxiliary fuel tanks, 59, 62. *See also* drop-tanks; escort fighters
AWPD-1, initial USAAF European air war plan, 62, 64, 65
AWPD-42, revised USAAF European air war plan, 64
Axis powers, World War II, 90, 162

"backchannel" communications, 23
Bacon, Francis (Sir), on historical truth, 16
Bacon, R. H. (Vice-Adm.), on battle cruisers' vulnerability, 52
Baghdad, damage in, in Gulf War, 136
Baldwin, Stanley (Rt. Hon.), on bombing threat, 57
Balkans, bomb damage in, in World War II, 141
balloon bombs, Japanese, 140
banzai charges, of Japanese Army, World War II, 8, 38
Barron, Frank, 114; quoted on creative individuals, 118
bases, military, costs of decontamination, 141–143
Bataan campaign (1941–42), 19; communications in, 81; as soldier's battle, 124; Death March, 161
battle: chaos of, 5, 11, 25, 48; historical evidence of, 2, 4, 21; information glut in, 23; morale in, 155; randomness in, 24
battle cruisers, doctrine and practice, 51–53; use of, despite diminishing returns, 69
battle drill, as doctrine, 104
battle dynamics: analogies of, 103; in Gulf War, 173, 177; models of, 157; order and randomness in, 153, 155; perception and description of difficult, 29; quantification of, 178; T. E. Lawrence on, 5; trends to fluidity in, 129; various models of 30, 178
battle management, of nuclear war, 161
battle police, 157
Bayesian analysis, 25
Bay of Pigs, Cuban invasion (1961), 148
bearing, as leadership trait, 102
Beatty, David (Admiral of the Fleet Earl), at Jutland, 52, 68. *See also* battle cruisers
Beer, Stafford (Sir), quoted, on confounding of plans, 88
behavior, human, as source of complexity-chaos, 155, 157. *See also* human factors; leaders; leadership; psychology
Belisarius, (Byzantine general), craftiness of, 120
bellometrics, bellometricians, 32, 97, 117
Berlin, Isaiah, quoted, on covering laws in history, 18
Berlin Wall, demolition of, 1989, 174
Betts, Richard, on surprise, 126

Beyerchen, Alan, quoted on von Clausewitz, 166
bias, of historians, 147
Bidwell, Shelford (Brig. Sir), 84
"big picture," in command-and-control process, 78, 152–153; and recall, 157
"Big Week" (1944), 67. *See also* Eighth Air Force
biological weapons, horizontal proliferation of, 141
Bismarck, German battleship, sinks *Hood*, 1941, 52
Black and Tans, in Irish Troubles, 53
black budget, 153
"Black Week," (October, 1943), 28. *See also* Eighth Air Force; escort fighters
Blackett, P. M. S., quoted on battle dynamics, 103
blame assignment, in military organizations, 82
blitzkrieg, 70
bluttkitt, 80
"body counts," in Vietnam War, 49, 138
Boer War (1899–1901), British army tactics in, 90, 159
Bohm, David, quoted, on creativity suppression, 121
Bolshevik Revolution, 1917, 7
Bomber Command, Royal Air Force, World War II, 63, 68
bombing, aerial, of cities, 161
Bonaparte, Napoleon (Emperor) 6, 20, 41, 42, 53; as 'Great Captain," 118; career cycle of, 164; on luck, 130; Russian campaign of, 47, 139; style of, 69, 115
bonding, of fighting units, 30, 81, 125
booby-traps, 101, 141
boundary conditions, in history, 19
bourgeoisie, in officer corps, 98
Braddock, Edward (Maj. Gen.), defeat of, 1754, 7, 86
Bradley, Omar Nelson (Gen. of the Army), opposes tank destroyers, 55; as "G.I. General," 80; style of, 102–103
Brandenburgers, German *Abwehr* special unit, 53
breakpoint, concept of, 102. *See also* panic
Bren gun carriers, 53
bridge failure, as complexity model, 92
brigs, naval prisons, 157
Bristol fighter, 60
Britain, Battle of (1940–41), 23, 109
British Army, 6; at El Alamein, 107; fractalizing of tactics, in Boer War, 90; indiscipline in, World War I, 78; El Alamein, 163; in North Africa, World War II, 8, 53–54; in Passchendaele offensive, 140; officers, style of, 98, 160; "Phantom" unit of, 88; tank tactics, World War I, 93, 99
British boarding schools, as rites of passage, 128
British Expeditionary Force, World War I, 139, 140; in Passchendaele campaign, 213. *See also* British Army; World War I
British Navy. *See* Royal Navy
Brusilov Offensive (1916), 107
buffalo, slaughter of, 139
Bulge, Battle of the (1944): as soldiers' battle, 124; Nazi assassination plot fear in, 80; surprise in, 43–44
Burckhardt, Jacob, quoted, on "irreplacability," 26
bureaucracy, ordering in, and historical evidence, 24; pressures in, and chaos-complexity, 154
Bush, George Herbert Walker (Pres.), and Gulf War, 43
butterfly wing (Lorenz) effect, 4, 10, 24, 170

Caesar. *See* Julius Caesar
Canadian corps, in 1918 offensive, 107
Cannae, Battle of (216 B.C.), 46
camouflage, in Cold War, 167; in Gulf War, 173; in World War I, 7, 101. *See also maskirovka*
Caporetto, Battle of (1917), as soldiers' battle, 124
Carden-Lloyd carriers, 53
Carlyle, Thomas, on "great men" in history, 18
Carlson, Evans (Brig.-Gen.), 8
Carnegie Foundation for Peace, reports of, 136
Casablanca Conference (1943), strategic bombing policy framed at, 63

Cassandra, 43, 114
casualties, in war: civilian vs. military, 136; declining tolerance for, in West, 79; effect on leaders, 88; in static warfare, 108; statistics of, 23
catalysis, of nuclear war, 167, 174
catastrophe theory, 11
causality and causation: in intelligence, military and diplomatic history, 150–151; in 1943 air battles, 68; in military history, 19, 25; training of historians and, 171–172
"c-cubed" technology, in Gulf War, criticized, 178
cellular structure, in intelligence systems, as complexity source, 149
The Censored War, 36
censorship, 21, 99, 128; in Gulf War, 136, 138
Central Intelligence Agency. *See* CIA
Central Pacific Theater of operations, World War II, 77
centralization-decentralization, debates over, 149, 152
Chamberlain, Neville (Prime Minister), misprediction of, 1940, 43
chaos: and war dynamics, 11–12; as characteristic of battle, 4–5, 76, 84, 87, 99, 156; as property of war, 126; as terminal state of war, 178; as theme in arts, 34; at Germantown, 26, 128; at Stalingrad, 91; coefficient of, 29; command process and, 178; computers and research in, 3, 71, 169–171; creativity and, 114–115, 120; descriptions of, 97, 123–124; distinct from complexity, 3, 12, 27, 160; diffusion of concepts of, 168–170, 178; doctrine and, 36, 70; enhances importance of details, 141; human factors, as source of, 82, 91; induction of, as tactical-strategic goal, 10, 92–94, 171; in Franco-Prussian War, 6; in German invasion of Russia, 1941, 165; in intelligence processes, 87, 145; in military history, 2, 4–5, 20; in World War I, 101; on Tarawa, 123; leaders' health, as source of, 161; New World Order, 168, 174; nuclear warfighting and, 166; panic, as source of, 102; order within, 109; policy processes and, 175; result of surprise attack, 10; technology and, 9, 171; transition to, from complexity, 171; war and torture, as forms of, 115
chaotic trajectory, in war, 116
charisma, of military leaders, 80
checklists, as form of doctrine, 104
chemical mortars, conversion of, World War II, 39
chemical weapons, 131, 136; horizontal proliferation of, 141
Cheney, Edward, quoted, on covering laws in history, 18
Chennault, Claire (Gen.), 57–58
chevauchee, 138, 139
China, 64, 91; bomber losses in, 1937–41, 43, 56, 58, 61, 146; floodings, 1938, 141; intelligence processes, in ancient, 146
China, Communist, armies of, 160; in Korean War, 164; use of camouflage, 167
Chinese Civil War (1945–49), 61
chivalry, in war, 159; in aerial war, 99
Churchill, Winston Leonard Spencer (Sir), 41, 116; as First Lord of the Admiralty, World War I, 51; at Casablanca Conference, 63; condemns science in war, 99; illnesses of, 161; Guildhall speech, 162; career of, 164; steadfastness of, 169
CIA (Central Intelligence Agency), U.S., purge of, 1970s, 151; sponsors drug and disease experiments, 1960s, 152
Civil War, American. *See* American Civil War
civilians, as casualties in war, 136; mass flights of, list, 110 n. 8
Clark Field, Japanese air attack on (1941), 164
classicism, and linearity, 166
Clausewitz. *See* von Clausewitz, Karl
clichés, military and naval, 158
Clinton, William (Pres.), on chaos, 34
close air support, in World War I, 7
closure need, of Americans, and chaos-complexity, 174
Coast Artillery Journal, Chennault articles in, 57

Coalition, casualties in forces of, Gulf War, 136
code-breaking, in Battle of the Atlantic, as source of chaos, 175
coherence, in command process, 108; paradoxes of, 156
cohesion, of fighting units, 30, 81, 124, 125
coincidence, and discovery in science, 129
cold war, 31, 41, 85, 151; complexities of, forgotten, 168, 174; complexity and superpower rivalry in, 167; Soviet strategic doctrine in, 166
Cole, Hugh, quoted, on "soldiers' battles," 36
"collateral damage," in Gulf War, 135–137; in Vietnam War, 137; portrayal in history, 91, 97, 135; in Gulf War, 207
colonial wars, 99
Colorado potato beetle, biological warfare vector, World Wars I and II, 140
combat: as chaos; 36, 76; historians of, 23; historical evidence of, 4 passim; intelligence, varying definitions of, 147; morphology of, 129; models of, human behavior abstracted in, 155; stresses, 1, 2, 101
combat processes, numerical analyses of, 118
combined arms tactics, 108; in World War II, 118
Combined Bomber Offensive, authorized at Casablanca Conference, 63; and 1943–44 air campaign, 56, 65–71
command discretion, 163
command error, 122
command and control: as source of chaos in war, 124; creativity in, 129; effect of computers on, 88, 98, 117, 156; effect on leadership, 80; and chaos in war, 159; and political authority, 177; and surprise and, 192; "ground truth—big picture"dichotomy, 153; in command process, 88; in Gulf War, 80, 81, 160, 173; in degraded states, 10, 49; in mechanized warfare, 47, 78; intelligence processes and, 147; in Vietnam War, 80–81; lack of a "good theory" of, 126; metaphors of, 105, 118, 130; Soviet logics of, 165–166; systems, 8–9; turbulence in, 48, 107
command information, varying definitions of, 147
command process: and chaos-complexity, 176; and maps, 79, 177; and surprise, 192; as source of chaos-complexity, 127, 163; coherence in, 48; creativity and, 129; dearth of descriptions of, 78–79, 82; effects of funding on, 133; graphic displays in, 77, 105; "ground truth" and, 152; in American Civil War, 160; metaphors of, 105, 118, 130; predictions and, 119; rivalry in, 69; semantics, 28, 126–128, 138, 155, 162; software in, 177; turbulence in 48, 50, 107
command psychology, 102
commanders: creativity and, 114, 130; idiosyncracies of, 88; irrationality of, 163–164; isolation from battle of, 100; judging prowess of, 118, 164
commanders, military: in battle, 19; potential effects of chaos-complexity theory on, 176
Commandos, British, in World War II, 53
command-staff process, rarely described in detail, 140, 160–161; commanders' motives and, 248
communications technologies, and battle dynamics, 88; as source of chaos, 78; non-linear concepts and, 3
compartmentation, effect on intelligence process, 149
complexity, 1, 26; and command and control design, 153; and sciences, 175; environmental damage and, 148; as function of numbers, 108; certainty in analysis of, 19, 106, 143 ; collateral damage and, 9; command prowess and, 167; communications technologies and, 88; creativity and, 130–135; definition of, vi; decision-making and, 1, 50; diffusion of theories, likelihood of, 168–170, 175–177; diffusion of concepts of, 169–171; distinct from chaos, 12, 27, 160; doctrine and, 36; doctrinal debate and, 104; effect on command and control, 130; effect of small details in, 24; environmental damage and, 148; human factors as source of, 148–49, 155–157;

imposing of on enemy, as goal, 93; increase in World War I, 101; in New World Order, 167–168, 174–185; intelligence processes and, 149; in Vietnam War, 136–38; in warfare, 3, 12; leaders' health as a source of, 161; leaders' view of, 158–59; military history and, 21, 23, 124; Soviet cybernetic theory and, 165; technology and, 177; uncertainty in analysis of, 19, 106; versus ordering in war, 50; technology and, 177; "what-iffing" and, 171; William Livingston on, 92
computers, effect of on battle dynamics, 153; command process, 130, 153,177, 178; chaos-complexity theoretics and, 3–4, 7, 169–171, 176; military operations and, 149; imaging of battle dynamics, 153; in complex organizations, 169; intelligence processes and, 98, 147–150; in Gulf War, 160, 173; limits of, 175, 177; "what-iffing" and, 171
Confederate Army (1861–65), casualties of, 9; historians of, 20; in Shenandoah Valley, 139
"convoy fighters," 61
convoys, naval, as tactical fractalization, 6; as tactical ordering, 100
"cookie cutter" models of nuclear war, 154
Corbett, Julian (Sir), 154; on battle cruisers, 51
CORONA HARVEST, U.S. Air Force air power study of, 89
"correlation of forces," Soviet doctrine of, 91, 166
coup d'oeil, 100
court-martial records, 157
covering laws, in history, 18, 142, 172, 172
Cozzens, James Gould, quoted, on disorder in military operations, 2
cran, 127
creativity: air tactics and, 68; analogies and models of, 120; and neurosis, 123; as function of surprise, 126; constraints on, in armed forces, 116; definition, 113–114; in intelligence processes, 113 passim; links with chaos-complexity, 130–131, 168; politics and procurement process, 131
Crimean War (1854–55), 81
crisis management, 31
Cromwell, Oliver (Lord Protector), 41; as prototypical dictator, 257
"crossing the T," tactical concept, 104
Cru, Jean Norton, quoted on staffs' distortions of fact, 78
cryptography, cryptanalysis: and non-linearity, 3; computers and, 98; in intelligence processes, 175
Cuba, 46, 47
cumulative effects, of environmental damage, 142–143
Custer, George Armstrong (Maj-Gen.), 20, 86; command style of, 102
customs of war, and war damage, 136; and World War II atrocities, 161
cybernetics, 117, 170; in Soviet military thought, 165
cycles, in war, 176

"Dam Busters," RAF attacks on German dams, 1943, 140
Darwin, Charles (Sir), 17, 170
D-Day (June 6, 1944), 26. *See* Operation OVERLORD
DDT, mass use of, World War II, 141
De Bono, Edward, 173; quoted on "blurry" and "sharp" brains, 118–119
deception, as source of chaos, 126; varieties of, in Cold War, 167
decision-making, rational models of, 150
decontamination, of military bases, 142–143
decryption, and command-and-control and intelligence processes, 141
defense analytics, and chaos-complexity, 11
defense policy, and chaos-complexity, 152–155, 166–168
defoliation, in Vietnam War, 137
de Gaulle, Charles Andre Joseph Marie (President), predictions of, 43; doctrines of, 125
de Pawlowski, G., quoted, on military history, 2
de Puy, William (Gen.), on clarity of tactical concepts, 39

Derfflinger, German battle cruiser, damaged at Jutland, 52
de Saxe, Hermann Maurice (Marshal Count), 5; versus linear tactics, 159
desert crust, destruction of, North Africa, World War II, 141
desert warfare, 171–173
de Seversky, Alexander P., proposes escort fighter development, 1938, 62
"devastated areas," World War I, 139
dialectic, Marxist, and complexity-chaos, 161; and Soviet military doctrine, 166
dialogue between "harmony and invention," 114, 129; in intelligence process, 145
dictators, 102; legitimacy and military roles, 102; modern, rise of, and technology, 81
diplomacy, and chaos-complexity, 10-11, 36, 150
discipline, and battle performance, 124; and military leadership, 80; in military history, 125. *See also* battle dynamics; command process; leadership; war, war dynamic, warfare
dispersion, effect of on combat dynamics, 109, 156, 159
dive-bombing, abandoned by U.S. Army Air Forces, 68
Dixon, Norman, 84, 114, 155
doctrine, military: alternatives, in shaping of, 40–41; and attractors, 49; as antithesis of effective stratagem, 50, 82, 145; as buffer to anxiety, 119; as dogma, 30, 35–37, 48–49, 83–84; battle cruiser doctrine, 50–53; debates over, 104; definitions, 34–38; effect of organizational dynamics on, 87–88; formats of, 28 passim; fragility in wartime, 35–37; history and shaping of, 3, 15, 24, 30, 47; sources of, 19; Soviet, 165–166; strategic bombing doctrine, 63–71; tank destroyer doctrine, 50–53
documents: and electronic transmission, 173; as basis of military history, 20–21, 157; inadequacies of, as evidence, 28, 77, 105, 162
Doenitz, Karl (Grosseadmiral), in Battle of the Atlantic, 100
Dogger Bank, Battle of (1915), 51–52
dogma, dogmatism; and doctrine, 30, 35–37, 48–49, 83–84
Doolittle, James (Lt.-Gen.), heads 8th Air Force, 67
double attractors, 49; conceptual dichotomies in military theory, analogous to, 104
Douhet, Giulio (Gen.), air power theories of, 57
Dowding, Hugh Caswell Tremenheere (Air Marshal Baron), copes with ambiguity, 106
Dreadnought, British battleship, 51
drop-tanks, 56, 58, 62–63, 65–67
"drum-and-trumpet" military history, 75
dual use, procurement policy, 154
du Guesclin, Bertrand (Count), 106; craftiness of, 120; evasive tactics of, 163; scorched earth tactics of, 139
du Picq, Ardant (Lt.Col.), 5
DuPuy, Trevor N. (Col.), on "timeless verities of combat," 104; Gulf War predictions of, 173
Eaker, Ira (Gen.), and 1943 stalemate, 65, 68; ordered to Mediterranean, 67; personality of, 69; proposes escort fighter, 1941, 63; quoted on escort fighter gap, 56; view of war, 106
East Africa, World War I in, 7
Easter Rebellion, Ireland (1916), 7
Eastern Bloc, in Cold War: collapse of unforeseen, 148; reliance on manpower, 167
Eastern Front, World War II. *See* Russian front, World War II
eccentricity, of leaders, 163
economics of war, complexity of, 169; paradoxes of, 153; and complexity, 169; uncertainties of planning, 34
Eder dam, RAF attack on, 1943, 140
"eddy currents of battle," 4
VIII Bomber Command, 63, 68
8th Air Force, U.S.: in bomber offensive, 1942–45, 56 passim; linearity of tactics, 69, 159; losses of, in 1943, 8, 66; 1943 setback analyzed, 8; operational problems of, 63, 65

Eighth Army, British, World War II, 173; and Gulf War battle plan, 173
88 millimeter gun (German), 56
Eisenhower, Dwight David (Gen. of the Army; Pres.), 2; quoted, on irrationality in decision-making, 89; pacific views of, 158; health of, 161; victories of, 164
El Alamein, Battle of (1942), as set-piece battle, 107; and Gulf War battle plan, 173
élan, 127
electrification: and command and control, 178; effect on warfare, 81, 87; on naval architecture, 50; modern dictators and, 81
El Guettar, Battle of (1943), tank destroyers in, 81
ELINT (electronic intelligence), 146
elite military units, 101
elites, capacity of, 170–171; measurement difficult, 118;
Ellis, John, 75; quoted, on incoherent impressions of war, 76
Emmons Board (1940), sets escort fighter priority, 72
encryption, in command-and-control and intelligence processes, 147
Engels, Friedrich, 18; and Soviet military doctrine, 166
Erickson, John, "technology-doctrine-style" triad of, 132 n. 10
environment, warfare's impact on, 135–143: as latent chaos, 143
escalation, nuclear, 174
escort fighters, 56–57, 61–67, 109
esprit de corps, 30, 80, 178
espionage: fictional portrayal of, 141, 150–151; mystique of, 146. *See also* CIA; intelligence processes
estimates of the situation, as form of doctrine, 104
Europe, 16, 15, 47, 99; fear of general war in, 167
European Theater of Operations, World War II, amphibious operations in, 90
evidence of war, historical: access to, control of, 148; deficiencies of, 162; inadequacy of, 23–25; unevenness of, 20–21
"expanding torrent," concept of J.F.C. Fuller, 104
expectations, confounding of, in history, 122
experience, and perception of war, 30; of military elites, 85–86
The Face of Battle, 89
Fabius Maximus, evasive tactics of, 106
fabrication, in history, 172
facsimile transmissions, and historical evidence, 23
Falkland Islands: Battle of (1915), 52; 1972 campaign, 160
fantasies, "what-iffing" in, 52
Fascism, 17. *See also* Nazism
fear, effect on thought of, 101
Federal Aviation (Howell) Committee, 1934, 58
Fenians, 19
"the Few," RAF fighter pilots, 1940, 23
fiction, portrayal of: war, 29, 30, 75, 77, 81, 91, 124, 166; intelligence processes, 146, 150–151; social class tensions, 159
field grade officers, command style of, 159
field manuals, 28; as doctrine form, 35
Field Service Regulations, German, 1933, 5
15th Air Force, opens Mediterranean air front, 96
"fighter belt," myth of, 64
fighter sweeps, of RAF, 1942, 63
fighting spirit, *versus* technology, debates over, 178
fingerspitzengefuhl, 100, 127
Finnish Air Force, 39
fire-and-maneuver, concept of, 104
fire exchange model, and complexity, 26
Fisher, John (Admiral of the Fleet Lord), as battle cruiser advocate, 51–52, 69
flamethrowers, as tactical surprise, 93
Flanders, drainage system of shattered, 1917, 6–7, 140. *See also* Passchendaele
floods, deliberate, World War II, 141
Flying magazine, 63, 158
"Flying Tigers," 57–58
Foch, Ferdinand (Marshal), 127

"fog of war," 2, 30, 69, 107
Ford Motor Company, 117; plant at Antwerp, 8th Air Force raid on, 64
Foreign Legion, French, 82
Foreign Legion, Spanish, 53
forests, as military targets, 139–140
forgeries, historical, 15
Forrest, Nathan Bedford (Brig.-Gen.), command style of, 108
"Fortress Europe," 64
Fouche, Joseph (Duke of Otranto), as spymaster, 152
Fourteenth Air Force, U.S., 57
fractals and fractalization, 170; as analogy of creativity, 131; definition of, vi; geometry of, 4; in modern warfare, 5–7, 88-89; in historical analysis, 149; in Napoleonic Wars, 6, 53; in World Wars, 100; of tactical forms, 6, 99; of British Army tactics in Boer War, 90
"fraggings," in Vietnam War, 160
France, 6, 46, 47, 55, 63, 81, 89, 125
Franco-Prussian War, 1870–71, 125; chaos in, 6
Frederick the Great, 102
free market economics, and chaos complexity, 152
French Air Force, heavy fighters of, 60
French Army: in Indochina, 167; tanks of, 53; offensive school of, 6, 46
French Foreign Legion, 82; officers' style, 160
French Revolution, Wars of, 6; armies of, 160; light forces in, 53
French, John (General Sir), 69
Freud, Sigmund, 17, 170; creativity of, 118; on morality as mask to vicious impulses, 162
friction, in war, 69; between command echelons, 79, 88; between social classes, 159; regional, 178
friendly fire, 122; in Gulf War, 177

Fuller, John Frederick Charles (Maj. Gen.), "expanding torrent" concept of, 104; quoted, on Boer War tactics, 90; on armored warfare goals, 93; on mechanization of armies, 99
Furmanov dictum, 100
Fussell, Paul, on censorship, 128
fuzzy logic/sets, 1, 26, 27, 68, 98, 154, 169, 170; and creativity, 114, 127, 130; and war dynamics, 168, 176

Gaddis, John Lewis, quoted, on inevitability and history, 150; on historians' continuity bias, 172
games: and doctrinal development, 31; theory of, 25
Gamelin, Maurice (Marshal), misappreciations of, 1940, 42
gangs, 158
Gellhorn, Martha, on censorship, 128
General Board, Artillery, 1946, disbands Tank Destroyer Corps, 56
general mission orders, 8–9. *See also auftragsbefehl/auftragstaktktik*
General Staff: German, war plans of, 1914, 46; Prussian, evolution of, 87; origins of, 8–9, 102; U.S. War Plans Division of, 50
General Motors Corporation, Antwerp, 8th Air Force raid on, 64
generalship, 59; as performing art, 115
General Systems Theory, 11, 170
general war, in central Europe, fear of, 167
Geneva, disarmament talks, 1930s, 53
Genghis Khan, 92, 116
German Air Force (Luftwaffe) auxiliary fuel tanks, used by, 87, 97, 99; heavy fighter development in, 89; 98; air defense tactics, 99–100; 100, 101, 157; long-range aircraft in Atlantic battles, 153
German Army (*Heer*): Field Service Regulations, 1933, 5; infighting in, in World War I, 69; infiltration tactics of, 7, 93; in North Africa, 1941–42, 54–55; in Russia, 1941–44, 55–56; losses of, 1870–71, 10; maneuver-based tactics of; officers' style, 98; Operation ALBERICH, 1917, 140; peace offensive of 1918, as soldiers' battle, 124; rivalry in officer corps, in World War I, 69; "surface and gap" doctrine of, 108
German Navy (*Kriegsmarine*), battle cruisers of, 52–58; submarine losses of, 6

Germantown, Battle of (1778), chaos in, 26, 128
Germany, 51, 55, 125; air defense of, World War II, 56, 63, 65, 66; bombing of, World War II, 99, 162; crops and forests of, as target, 139–140; dams attacked in, 1943, 140; RAF bomber losses over, 109. *See also* Franco-Prussian War; General Staff; German Air Force; German Army; German Navy; Napoleonic Wars; World War I; World War II
The Gestalts of War, 36
Gettysburg, Battle of (1863), 127
GHQ Liaison Regiment, of British Army. *See* "Phantom"
GHQ Air Force (U.S. Army Air Corps), 59
"ghost of Napoleon," 121
Giap, Vo Nguyen (Gen.), evasive tactics of, 106
Gilmore, J. Bernard, quoted, 12
Gödel's theorem, 77
Goering, Hermann (*Reichsmarschall*), in Battle of Britain, 39
Golden Horde, 26
Grand Fleet, British, at Jutland, 52
Grandmaison, Henri (Col.), 28
grand strategy, 36
Grant, Ulysses Simpson (Pres.), as outsider, 121; command style of, 83, 102; idiosyncracies of, 123; victories of, 164
graphic display, of command process, 177
Gravelotte, Battle of (1870), 10
Great Britain, 8, 47, 117, 158; as source of drop-tanks, 65; in Malaya, 167; in Suez affair, 161
"Great Captains," 118, 120
Great Plains, buffalo slaughter on, 139
Greene, Nathaneal (Maj.-Gen.), 121
Gregory, R.L., quoted, on imperceptibility of complex phenomena, 103
"ground truth," in command and control, 152–153
Guderian, Heinz (Gen.), 104; command style of, 109
Guernica, bombing of (1937), 135
guerre d'course, 104
guerrilla warfare, 99, 109. *See also* low-intensity conflict; small wars
Guildhall speech, of Winston Churchill, 1941, 162
Gulf War (1991), 174–175, 177; command and control in, 80, 81, 160; bombing in, 109, 206; environmental effects of, 135, 136, 143; intelligence problems in, 147; deception in, 167; pundits' predictions confounded in, 42
gun cameras, as historical evidence, 156
"gun-line," in 1982 Falklands campaign, 160

Hackett, John (General Sir), 171
Hackworth, David (Col.), on warrior spirit, 104
Haig, Douglas (Field Marshal Earl), 139; command style of, 83; displaces Lord French, 69; victories of, 164; view of war, 106
Halsey, William F. (Fleet Adm.), command style of, 83
Hamiltonian system, as analogy of military forces, 13 n. 7
Hannibal, 92, 115; as "Great Captain," 118; tactics of, in Second Punic War, 46
Harris, Arthur (Air Marshal Baron), 68; threats in broadcasts to Germany, 162
hauptschlacht, in Central Europe, 167
Hawker, British aircraft firm, 61
Hawking, Stephen, on "good theory," 126
Hayek, Theodor, opposes economic regulation, 169
hazing, and creativity, 115
"headquarters syndrome," 85
health, of leaders, as source of complexity, 161
heavy fighters, development of and role in World War II, 56–71
Hegel, Georg Wilhelm Friedrich, 169; idealism of, 166, 170
Heinkel, German aircraft firm, 60
Heisenberg, Werner, Uncertainty Principle of, 130
helicopters, commanders' use in Vietnam, 178
Heligoland Bight, naval clash in, 1914, 52

hemispheric defense, U.S. doctrine, 1937–41, 58
Henderson, G.F.R. (Gen.), quoted, 12
Henschel, German aircraft firm, 60
hermeneutics, 33 n. 7
heroism, paradoxes of, 137
Hess, Rudolf (Reichsfuhrer), mental state of, 161
Heydrich, Reinhard (*Obergruppenfuehrer*), assassinated, 1942, 80
High Seas Fleet, German, at Jutland, 52
Hindenburg Line, 140
hindsight, and history, 20, 69, 77, 149
Hinsley, F. H., quoted, on types of wars, 92
Hiroshima, atomic bomb attack on (1945), 142, 158, 166
historians: and analysis of war, 1, passim; 66; and chaos-complexity theory, 11, 153; and propaganda, 15–16; causation, and training of, 150, 171; combat, 23; intelligence and, 150–152; use of personal accounts of war, 135; use of documents, 137
historical truth, elusiveness of, 148
historicism, 16, 17
history, historical analysis: ambiguities in, 107; analytical problems in, 47, 89–92, 121, 150, 153; 163, and planning, 3; as basis of forming doctrine, 15, 27, 28, 29, 31, 34, 40–41, 46, 120, 154; as retrospective intelligence, 148; battle dynamics portrayed in, 75, 92, 97; chaos-complexity theory and, 171–172; covering laws in, 18, 142, 172, 178; deficiencies of, 2; distortion of, 23, 30, 31; fabrication in, 172; intelligence, portrayal of in, 225; laws of/in, 66, 166; non-linearity and, 32, 145 passim; of science and technology, inventions and accidents in, 129; patterns in, 19; psychological factors in, 163; search for lessons in, 20; secrecy, and, 162; subtleties affecting, 98; war damage, portrayal in, 136
Hitler, Adolf, 41, 42, 116, 170 61, 62, 258; as failed visionary, 43; in Munich crisis, 1938, 47; avoids wounded, 83; bridles General Staff, 102; ill health of, 161
Ho Chi Minh, ill health of, 161
Holland, flooding of, 1944–45, 141
Holley, I. B., definition of doctrine, 37
Holocaust (1941–1945), 161
Homer, 43, 114
Hood, British battle cruiser, sunk, 1941, 52
Hooker, Thomas (Maj. Gen.), defeated, 106
Horatius, at the bridge, 69, 169
horizontal escalation, in New World Order, as source of chaos, 174
horizontal proliferation, of nuclear, biological and chemical weapons, 141
horrors of combat, 76, 128
Hughes, Thomas (Vice-Adm.), quoted, 38
human factors, as source of complexity, 162–163; behavior abstracted in games and models, 155. *See also* behavior; idiosyncracy; psychology
HUMINT (human means of gathering intelligence), 146; fluctuating fashionability of, 151
Hundred Years' War (1337–1457), 139
Huns, onslaughts of, 129, 139
Hussein, Saddam, 43, 102

ICBM (intercontinental ballistic missile), Soviet, throw-weight of, 166
Iliad, 43, 83, 114
Inchon, landing at (1950), planning of, 86
indices, operational, scarcity of, 89, 103
"indirect approach," concept of, 107
indiscipline: in battle, 155; in World War I, 139; of military leaders', 125–126, and instances listed, 133 n. 31
individual behavior, as source of chaos-complexity, 149, 155, 157, 160. *See also* human factors; idiosyncracies; leaders, leadership; psychology
Indochina, losses in wars in, 10; France in, 167
inevitability, in history, 171–172
infantry tactics, in Korean War, 106
infiltration tactics, German, World War I, 7, 93, 99
Inflexible, British battle cruiser, 52
informal processes, and history, 172; and intelligence, 152; in battle, 156

initial conditions, sensitivity of pre-chaotic systems to, 143
insurgency, 99, 109
intelligence analysis. See intelligence processes
intelligence community, U.S., as source of creativity research, 113
intelligence, human, pejorative terms for, 134 n. 36
intelligence processes, creativity and chaos-complexity, 146–152; fictional views of, 146, 150–151; historical perspectives on, 15, 66, 148; Gulf War, problems with, in, 147; Law of Location and, 149; mystique of, 150; popular portrayals of, 146–147; international law, and war damage, 136; and World War II atrocities, 161
invention, processes of, 152
Invincible, British battle cruiser, 51; lost at Jutland, 52, 75, 76
Iran-Contra affair (1981–86), 148
Iran-Iraq War (1980–88, losses in, 10; frontal assaults in, 107
Iraq, invasion of Kuwait by, 1990, 135, 148; civilian casualties in, 136
Iraqi armed forces, operations in Gulf War, 173
Ireland, "Troubles" in (1916–22), 7, 53
irrationality, and creativity, 122, 130; in command process, 164
Israeli armed forces, officer corps origins, 86
Italian campaign (1943–45), tank destroyers in, 54–55
Italian Army, light tanks of, 53; in North Africa, 1940–42, 54
Italian Navy, light units of, in World War II, 51, 70
Italian realist films, 136
Italy, U.S. bomber bases in, in World War II, 64

Jackson, Thomas (Maj. Gen.) ["Stonewall"], command style of, 84, 108, 115; idiosyncracies of, 123; in Shenandoah Valley, 164
Japan, airborne incendiary balloon attacks of, on Pacific Northwest, 140; air war against, 109; armed forces, *kamikaze* units of, 23
Japanese Army: air forces of, auxiliary fuels used by, 59; *banzai* charges of, 8; officers' style, 98; heavy fighter development of, 61
Japanese Navy, 20, 39
Jena, Battle of (1806), aftermath of, 99
jet aircraft, German, 56
"Jock Columns," North Africa, 1941, 53
Johnson, Lyndon (Pres.), as war controller, 116; ill health of, 161
Jomini, Antoine Henri (Baron), 103, 177; on unclear rules in amphibious warfare, 110 n. 9
Jones, John Paul (Capt.), 103
Jones, R.V., quoted, on unreliability of witnesses, 117
journals, as record of military operations, 105
journalists, as judges of military operations, 164; in Gulf War, 178; policymakers view of, 150;
Julius Caesar, creative style of, 115
junior officers, demeanor of, 159
Jutland, Battle of (1916), 50, 52, 68, 69

Kahn, Herman, department store metaphor of war of, 38; on unstable nuclear balance, 111 n. 37
kamikazes, 23, 70
Kampfzerstorer, German Air Force heavy fighter concept, 60
Katyn Forest massacre (1940), 135
Keegan, John, on evidence of war, 22; on commanders' concern for panic, 102; quoted, on battle management, 89
Khwarizmian Empire, destruction of, by Mongols, 138
Kiggell, Launcelot (Lt.-Gen. Sir), quote, on Passchendaele, 6–7
"kill boxes," in Gulf War, 173
Kimmel, Husband (Adm.), 43
King, Ernest Joseph (Fleet Adm.), mispredictions of, 43
King Ranch chain, use of in Vietnam, 137
kings, as martial leaders, 101–102
kleine kriege, 7

Koch, Noel, quote, on disorder as terrorist goal, 93
Konev, Ivan (Marshal), quoted, on history of World War II, 75
Korean War (1950–53), 67; air tactics in, 109; communications in, 109; bonding and cohesion concerns, following, 125; Chinese intervention, mispredictions and, 44; constraints on use of force in, 254 infantry tactics in, 106, 107; linearity of tactics in, 137, 159; MacArthur seeks to widen, 158
Kosko, Burt, quoted, on theory-reality difference, 149
Kuhn, Thomas, on "paradigm shift," 11, 106
Kuwait, oil field fires in, 1991, 135, 136; invaded by Iraq, 1989, 148

Lanchester, F. W., "square law" of, 26
land mines, 141
latitude, of commanders, 164
Law of Location, 108; in intelligence process, 149
Law of Requisite Variety, 26
Law of the Situation, 9, 84, 125, 129, 147
laws, in/of history, 18, 66, 142, 172, 178; of war, 159, 176. *See also* Law of the Situation
Lawrence, T.E. (of Arabia), 158; as outsider, 121; quoted, on irrationality in war, 5; views of war, 157
leaders: career cycles of, 164; chaos-complexity and, 31, 158, 161, 175; creativity and, 114–130; experiences of, 85–86; health of, as source of complexity, 161; idiosyncracy, 122–123, 163–164; in American Civil War, 160; indiscipline of, 125; influence on doctrine, 70; mystique of, 104–105; overidentification, with followers, 81; prediction and, 42–44; rhetoric of, Allied, in World War, 162, 170; role in intelligence processes, 149–152; stresses on, 101; tenacity of, 69, 125
leadership: chaos-complexity and, 40, 81, 93, 94, 96, 97; as opposed to command authority, 156; command process, 30, 48, 49; discipline and, 160; immeasurability of, 163–164; mystique of competence, 105–106; styles, and, 60, 80–108 passim; versus command, 156
Lee, Robert E. (Lt.-Gen.), command style of, 102, 169; career cycle of, 164; defeat and, 106. *See also* American Civil War; Army of Northern Virginia
LeMay, Curtis (Gen.), devises "box" defense tactics, 65; bellicose views of, 158
Lend-Lease, 54
Lenin, V. I. (Ulyanov), interest in von Clausewitz, 37, 166; faith in historical laws, 166
lessons of war, search for, in history, 20, 30, 47
"lethal density," RAF concept of, 60
liberalism, 27
Libya, British tank losses in, World War II, 8
Liddell Hart, Basil (Sir), 104, 107
light forces, in Napoleonic era, 53; in World War II, 51
"light-heavy" debate, in military theory and history, 159
light infantry, 11; in 18th century, 8, 129
light tanks, 53–54
limited wars, 159, 167
Lincoln, Abraham (Pres.), 4, 116
linearity, of tactics, 5–8: and battle dynamics, 153; basis of, 8, 28–29; chaos in war and, 98, 166; historical analysis and, 146; in aerial warfare, 109; in Gulf War, 173; in various wars, 107, 129; in Vietnam, 137; in military ceremonies, 106; 97; policy sciences and, 175
Livingston, William, quoted, on complexity, 92
logs, message, as record of military operations, 105
"logical incrementalism," 40
logistics, confusion of, in DESERT STORM, 173; in military history, relative coherence versus combat description, 108; various models of, 154
Lorenz, Edward, complexity theories of, 3–4, 10, 24, 170

Loftus, Elizabeth, on memory deficiencies, 22
losses, in battle, declining tolerance for in West, 79. *See also* casualties
long-range aircraft, in World War II sea warfare, 100. *See also* Doenitz, Karl; German Air Force; German Navy; U-boats
lower participants, effect on complex dynamics, 162; theory of, 105
low-intensity conflict, 167; in New World Order, 168. *See also* guerrilla warfare; small wars
Ludendorff, Erich (Gen.), view of war, 106, 158
Luftwaffe. *See* German Air Force
Lundy's Lane, Battle of (1814), 20
Lutzow, German battle cruiser, damaged at Jutland, 52
Lyon, A. J. (Brig.-Gen.), proposes "flying battleship," 66

MacArthur, Douglas (Gen. of the Army), career cycle of, 158, 164; exiled by Pershing, 69; predictions of, 64, 166; advocates PT boats, 51
Macaulay, Thomas Babington (Lord), 31
machine guns, effect on warfare of, 100
McCloskey, David, quoted, on norming tides in history, 172
McKinley, William (Pres.), 116
McKnight, Frank H., quoted, on planning uncertainties, 34
McNamara, Robert S. (Sec. of Def.), quantification of military planning by, 117, 137
Maginot Line, 47
Magna Carta, 16
Malayan Insurgency, 1848–1962: British in, 167; rubber trees slashed, 141
Malmedy massacre (1944), 161
Malta, bombers losses over, 1941–42, 109
managerial revolution, hopes of confounded, 98
Mandelbrot, Benoit, and fractal geometry, 3–4
maneuver warfare concepts, 104, 173
Mann, Steven R., 36
Mansfield, Sue, 36
Mao Ze Dong (Premier), 41; interest in von Clausewitz, 166
maps, military, 79; as image of war, 105; limitations of, as evidence, 177
March to the Sea [Sherman's] (1864), 137, 139
Marcus, John, on historicity, 17
Marder, Arthur, quoted on judging battle results, 87
Marine Corps. See United States—Marine Corps
Marlborough, Duke of (John Churchill), command style of, 80; victories of, 164
Maritain, Jacques, quote, on "vectorial laws," 18
"Maritime Strategy," U.S. Navy, 1980s, 70
Marshall, S. L. A., quoted, on battle dynamics, 4, 22
mathematics, application in military operations, 117
Marx, Karl, 17, 41, 170; in Soviet military doctrine, and faith in historical laws, 166; on revolutions as "locomotives of history," 169
Marxism, Marxists, 17, 18, 169
Marxist-Leninism philosophy, and command and control, 165
maskirovka, 165
Mata Hari, 146, 158
Materiel Command, USAAF, 63; drop-tank procurement snarls, 65
Maxwell, James Clerk, 130
Meade, George (Maj. Gen.), 102
Mechanic, David, theory of lower participants of, 105
mechanized warfare, environmental effects of, 141; "expanding torrent" concept of, 104; mechanization, effects on warfare of, 8
media, news, Gulf War focus of, 155
Mediterranean Theater of Operations, amphibious operations in, World War II, 90
memoirs, portrayal of war in, 89–91
memory, limits of, 175; stresses of combat and, 157
message forms, as a form of doctrine, 104
Messerschmitt, German aircraft firm, 60, 66

metallurgical revolution, 50
metaphors: of chaos-complexity, 97, 166, 170, 178; of Cold War, 167; of command process, 3, 126–127, 176; of history, 172; of intelligence process, 148–149; of New World Order, 168
micromanagement, 79, 81
Middle East War (1973), losses in, 10
Midway, Battle of (1942), 20; Japanese war gaming of, 39
miles gloriosus, 82, 159
militarism, 17
military bases, decontamination of, 141
military historians: causation traced by, 24, 25, 47; command performance judged by, 164; military experience of, 76; perceptions of war of, 91, 125
military-historical science, Soviet concept of, 2
military history: analogous to mass disaster report, 176; and "lessons of war," 31; battle portrayal in, difficulties of, 75–76, 88, 91, 108; causation in, 47, 116; chaos-complexity and, 3–4, 172, 178; distortions of, 23–24, 82; doctrine formulation and, 3, 15, 24, 30, 47; double attractors in, 104; evidence in, 1, 2, 20–21 and passim; law of location, and writing of, 108; maps and, 177; military professionals' use of, 15 passim, 160; non-linear perspectives on, 19; ordering in, 176; posturing in, 176; veterans' skepticism regarding, 2, 21, 30, 31; "what-iffing" in, 171
military law, and war damage, 136
"military mind," 57, 71; and creativity, 174
military operations, effects on environment, 136–143
military organizations, constraints on creativity in, 158
military police, 157
military professionals, debates among, on technology versus fighting spirit; use of military history by, 15; implications of chaos-complexity for, 153–154, 174; views on war and order, 158, 177
military systems, analogous to articulated machinery, 107
military theories, 106; and turbulence, 166; abstraction of human behavior in, 155
mind-set, of Western elites, and chaos-complexity, 174
mines, land, 101
misestimates of military operational effects, 47–49
mission-type orders, 70. *See also auftragsbefehl/ auftragstaktik*
Mitchell, William ("Billy") (Brig.-Gen.), forest fire plan of, 140; predictions of, 43; proposes "convoy" escort fighter, 61
models: of battle dynamics, 102; of chaos-complexity, 3–4, 154, 175
Mohne dam, RAF attack on, 1943, 140
Moltke. *See* von Moltke, Helmuth
Monash, Sir John (Gen.), command style of, 108; quoted, on military narrowness, 115
Mongols, tactics, as model for tank warfare, 47; destroy Khwarizmian canals, 138
Monroe Doctrine, 37
Monte Carlo processes, 11
Montgomery, Bernard Law (Field Marshal Viscount), 146; command style of, 83; idiosyncracies of, 123; reputation of, 164
Montrose, James Graham (Marquis), craftiness of, 120
Moore, John (Maj. Gen. Sir), light infantry system of, 8
morale, 125
Morgan, Daniel (Brig. Gen.), at Cowpens, 164
morphology, of combat, 129
Mosby, John (Col.), 139
Moscow, German air raids on, World War II, 61; burned, 1812, 211
motion pictures, battle images in, 21, 77
motives, human, and chaos-complexity, 91, 162–163; and history, 42
motor torpedo boats, 50–51
Muhammad, 41
Mukden, Battle of (1904–1905), 10
Munich crisis, 1938, 47
Murphy, Audie (Capt.), 157
mutinies, in World War I, 7, 160

Nagasaki, atomic bomb attack on, (1945), 142
napalm, use of, in Vietnam War, 137, 161; in World War II, 161
Napier, John, and conceptual symmetry, 169
Napoleon I. *See* Bonaparte, Napoleon
Napoleon III, 9, 116
Napoleonic Wars, 41, 159; historical momentum of, 99; light and heavy forces in, 6
narrative form, of historical description, 91
National Command Authority (NCA), 161
Nazis, 15, 17, 18, 23, 43, 54, 55, 166; war criminals tried, 161
Nelson, Horatio (Admiral of the Fleet Lord), 80, 118; idiosyncracies of, 123
nettoyers, of French Army, World War I, 53
New Guinea, tribal wars in, 129
Newman, John (Cardinal), quoted, on dogma, 35
New Military History, 75, 105
New Model Army, 160
New Orleans, Battle of (1815), 9
Newton, Isaac (Sir), 130, 152, 170; symmetry in theories of, 169
New World Order, 34; complexities of, 168, 174–175
Nixon, Richard Milhous (Pres.), in Watergate crisis, 161
NKVD, units of, in Red Army, 165
non-commissioned officers, bellicosity of, 158
non-linearity, concepts of: and battle forms, 30; computers and, 98, 173, 176; effect on military affairs, 3; implications of, 25; and military history, 105–106; Russian preeminence in, 165; diffusion of, 168–171
Normandy campaign (1944), 118; tank destroyers in, 54–55; breakout, and Allied Fighter range, 67
North Africa, U.S. bomber bases in, 64; end of blitzkrieg in, 90–91
North African campaign (1941–42), tank destroyers in, 81–82; tank combat in, 90; chivalry in, 99; Special Air Service in, 109; desert crust destruction in, 141
North Korea, armed forces, deception techniques of, 173
North Vietnam, armed forces, deception techniques of, 173
Nuclear Age, in small wars, 7
nuclear war-fighting: battle management networks, 161; chaotic potential of, 166
nuclear weapons, 117, 131, 136; atmospheric tests of, 142; environmental effects of, 142; horizontal proliferation of, 141. *See also* Cold War; thermonuclearetics
numerical analyses, in military theory, 6, 117; in Vietnam War, 137
Nürnberg Tribunals (1946–50), 47

objectivity, in historical analysis, 147
officers, commissioned, and social class, 159
official histories, screening of, 147
oil spills, from sunken ships, World War II, 140
"Old Breed," U.S. Marine Corps, World War II, losses of, 23
Omdurman, Battle of, 1896, as firepower triumph, 7; slaughter at, 10
Operation ALBERICH (1917), 140
Operation DESERT STORM (1991); 172–173, 177; losses in, 10, 140. *See also* Gulf War
Operation NEPTUNE (1944), 128
Operation OVERLORD (1944), 128
Operation POINTBLANK (1943–45), 63
operational art, 24, 188; and surprise, 122; of Red Army, 165–166
operational data, dearth of, 117
operational research, 11, 97, 98, 111
operational thinking, Soviet concept of, 122
O'Quinn, James Brian, quoted, on "logical incrementalism," 40
Oradour-sur-Glane, massacre by S.S. at (1944), 135
ordering: commanders' impulse to, 81, 107, 158; human impulse to, 6, 108–109; in defense planning, 155; in Gulf War, 160; in history, 19, 24, 48, 128, 170; search for, in history, 104; versus creativity, in armed forces, 155

orders, military, as form of doctrine, 104
Organs of State Security, Russia/ Soviet Union, and dialectic, 165
overidentification, of leaders with followers, 81, 82

Pacific Northwest, forests of, as Japanese balloon bomb target, 140; spruce stands devastated, 1917–18, 139
Pacific Theater of Operations, World War II, 39, 43; amphibious warfare in, 90; armored forces in, 54; DDT use in, 141; PT boats in, 51
pacifists, pacifism, 97; after World War I, 100
Panama, invasion of, 1989, planning-execution gaps in, 177
panic, in battle, 102, 155
paradigm shift, chaos-complexity theory as a, 11, 107, 170
Paris gun, as strategic surprise, World War I, 93
Parkinson's law, 119
paradoxes: in economics of war, 153; in post-Cold War era, 130, 156–164
Passchendaele, Battle of, 1917–18, 6, 140
patterns, orderly, search for: confounded in war, 131; in conceptualization, 130; within apparent randomness, 198
Path Finder Force, Royal Air Force, 1942–45, 39
Patton, George S., Jr. (Gen.), 2; antitank doctrine of, 55; command style of, 83, 102; eccentricity of, 84, 123, 158
Peace Offensive, 1918, of German Army, as soldiers' battle, 124
Pearl Harbor (1941), 7, 57, 59, 64, 68; predictive failures and, 42, 43
Peat, F. David, quoted, on creativity suppression, 121
People's Republic of China. *See* China—Communist
perception, human: discontinuity among echelons, 48–49; effects of combat on, 2, 22; limits of in sensing complexity and chaos, 3, 12, 68, 70, 92, 103, 175, 177; in writing history, 172
Pershing, John J. (Gen. of the Armies), "exiles" MacArthur, 69
personality, role in sustaining, doctrine, 69
Petain, Henri Philippe Omer (Marshal of France/ Premier), 102
"Phantom," unit of British Army, 88
Philippines campaign, 1941–42, tank destroyers in, 54
photogrammetry, and intelligence process, 147
Pierce, Charles Saunders, quoted, on "irritation of doubt," 146
"pillar of fire by night," Liddell Hart's concept of, 104
Pilsudski, Josef (Marshal), 102
Plains Indians, 139
planning, plans, military: 3; as buffer to anxiety, 119; confounding of, in war, 88, 122; historical basis of, 120; in Gulf War, 173
"pocket battleship" fighter aircraft concept, 62
poison gas, 68, 93, 132 n. 5, 140; surprise potential of, lost, in World War I, 102
polemics, of leaders, as source of complexity, 162–163; and history, 167
polemologists, polemology, 97
policy process: analysis of, 122; and chaos-complexity, 175 passim; and doctrine, 35; discontinuity and irrationality in, 122, 149, 227; failure of, 122; linear bias in, 175–176; link to doctrine, 35
popular culture, influence on publics and elites, 78; images of soldiers in, 82; images of war in, 31, 51
post-traumatic stress disorder, 30
potato beetles, Colorado, as biological warfare vector, 140
potok (information flood), Soviet concept, 23, 165
Powell, Enoch (M.P.), quoted, on human foibles in policy process, 149
Pratt, Fletcher, 138; on military secrecy, 138
prediction: and command process, 119–120, 129; lack of index of effect, 43–44; leaders' claims to special ability in, 181; failures in, from perspective of chaos-complexity, 222.
"predictive hopelessness," 169

principles of command, lack of, 120
principles of war, 26, 104, 120, 127; failure of, as a predictive aid, 176; list of, 111 n. 20
printing press, moveable type, effect on propaganda of, 15; in history, 16
prisons, 158
procurement process, effects of funding on, 153–154
propaganda, as complexity increment, 178; in Vietnam War, 138; rise of, 15–16, 80; Soviet, in World War II, 165. *See also* media; psychological warfare
provinces of war, Clausewitzian concept of, 130
Prussian Army, 6; command logic of, 8–9; General Staff of, origins, 102; images of officers, 82; reforms in, 1807–1813, 8
psychological warfare, 80, 93, 138
psychological measurement, in armed forces, as complexity source, 80
psychology, and command, 83, 101, 130
The Psychology of Military Incompetence, 64, 114, 155
PT boats, 51
Punic War, Second, 109
"pure science," myth of, 131
Pyle, Ernie, quoted, on confusion of war, 76

Qaddafi, Muammar, 156, 180 n.22
quantification, of battle dynamics, 117
Queen Mary, British battle cruiser, sunk at Jutland, 52
Q-ships, surprise potential of lost, 68, 132 n. 5; as fractalizing of tactics, 100

radar, in antisubmarine warfare, World War II, 100; in command and control, 177
Radford, Arthur (Adm.), bellicosity of, 158
radios, effect on historical evidence of war, 23
railways, in warfare, 8, 177
randomness, 1; as opposed to chaos, 3; depicted in fiction, 29; in policy process, 31
Rangers, U.S. Army, in World War II, 53
rational actor models, of leadership, 50, 82; of intelligence process, 150
rationalism, 24
rationality, and command process, 163; in command process and policy analysis, 122; *mythos* of, 176; versus irrationality, 130
Reagan, Ronald Wilson (Pres.), shooting of, 161; doctrine of, 37
reality, conceptual irreducibility of, 175
Red Air Force, heavy fighter aircraft in, 61
Red Army, assault guns of, 55; command and control in, World War II, 165; confusion of battle, described, 89–90
"red phone," and U.S. presidency, 161
reflex control, Soviet concept of, 93
Reform Movement, U.S., post-Vietnam war, 158; command and control critiques of, 178
regional friction, as complexity source, 58
Reichswehr, tactical doctrine of, 94, 108
Renaissance, 98, 117; concepts of, as models, 171
Rennenkampf. *See* von Rennenkampf
Republic Aviation Corporation, proposes auxiliary fuel tanks, 62
research and development, military, and creativity, 131
Rhine river crossing, 1945, 107
Richardson, F. M. (Brig.), 84
Richardson, Lewis F., boundary density-conflict theories of, 108, 174
Richtofen. *See* von Richtofen, Wolfram
rites of passage, military: and behavior, 156; linearity and, 106, 159; effect on creativity, 115, 128
ritual warfare, 129. *See also* tribal warfare
rocket aircraft, German, World War II, 106
rockets, 131
Roeder, George, quoted, on complexity in World War II, 36
role-play, leadership as, 69
Rome, 46, 85
Romans: auguries of, 152; *miles gloriosus* of, 82, 159; razing of Carthage by, 138; view of war as chaos, 94
Romanticism, 16, 166
Rome plow, used in Vietnam war, 137
Rommel, Erwin (Field Marshal), 108; as "Great Captain," 118; as outsider, 121;

British commando raid on, 80; on confusion in battle, 89; quoted, on battle dynamics, 78, and on military theories, 104
Roosevelt, Franklin Delano (Pres.), 41, 43; at Casablanca Conference, 63; declares war, 1941, 162; ill health of, 161
Rosenberg, Alfred, 18
"round-the-clock" bombing. *See also* Combined Bomber Offensive
Royal Air Force: B-17 losses of, 58; bomber tactics problems, World War II, 40; escort fighter views of, 58, 62; environmental damage by, World War II, 140; fighter pilots of, in Battle of Britain, 1940, 68, 162; fighters escort U.S. bombers, 63; heavy fighter development of, 60–61; Path Finder Force of, 39
Royal Navy, 38; aircraft carriers in, 51–53; battle cruisers of, 51–53
Ruelle, David, quoted on turbulence as graveyard of theory, 37
rules of engagement, and war damage, 136
Russia, 46; armies of, 160; German invasion of, 1941, 166; Napoleon's invasion of, 1812, 139
Russian Army, 8; command and control logic in, 44 n. 3; 103; military doctrine of, 118
Russian Front, World War II, winder of 1941, 47; soldiers' battles on, 189
Russo-Japanese War (1904–05), 107

Samsonov, Aleksandr Vasilyevich (General), rivalry with von Rennenkampf, 70
samurai, 104
Santiago, Battle of (1898), 127
satellites, orbiting, and command and control, 177; and intelligence process, 146, 151
Savannah, fall of, 1864, 139
Savonarola, Giralamo, as failed visionary, 43
Schell, Jonathan, "nuclear winter" vision of, 141
Schley, Winfield Scott (Commodore), as inadvertent victor, 106
Schlieffenplan, failure of, 9, 41, 46. *See also* von Schlieffen, Alfred; World War I
schwerpunkt, concept of, 90, 127
science-fiction, "what-iffing" in, 171
science, history of, 129
sciences, application of, in war: and chaos-complexity, 175; 117, 131
Second Front, demands for, 1942, 63
2nd Raider Battalion, U.S. Marine Corps, 8
secrecy, effects on historical analysis, 162, 172
secret weapons, use of, in World War I, 93
selbstandigkeit, 9
semantics, as buffer of violence, 28, 126–128, 138; and military documents, 162–162
sensitivity to initial conditions, xi; and cumulative environmental damage, 143; in military history, 178. *See also* subtle forces
Serapis, British frigate, 103
7th Armored Division, German, Rommel commands, 1940, 89
Seversky Aircraft Company, 62
sexuality, in policy processes, 149; war and violence, and, 158
Seydlitz, German battle cruiser, damaged at Jutland, 52
Shafter, William (Maj.-Gen.), 47
"shaking down," naval practice of, 106
Shenandoah Valley, Sheridan devastates, 1864, 139
Shannon, Claude, 26, 50
Sheridan, Philip Henry (Maj.-Gen.), favors buffalo slaughter, 139; devastates Shenandoah Valley, 1864, 139
Sherman, William Tecumseh (Maj.-Gen.), 186; idiosyncracies of, 122–123; leads March to the Sea, 159
Sherrod, Robert, on Tarawa chaos, 123
Shiloh, Battle of (1862), 20
Short, Walter C. (Gen.), misappreciations of, before Pearl Harbor, 42
Sicilian campaign (1943), tank destroyers in, 55
Sicily, U.S. bombers based in, World War II, 64
SIGINT (signals intelligence), 146
signal-to-noise ratio, as model for chaos co-efficient, 12, 27

simplicity: and complexity, 130, 166–167; as principle of war, 26, 127, 155; in intelligence process, 152; Western elites' preference for, 174–175; von Clausewitz on, 107
simulations: command and control, and, 152; degrees of reality in, 49; doctrinal development and, 31
Singapore campaign (1941–42), 19
Sino-Japanese War (1937–45), scorched earth tactics in, 141; unescorted bomber losses in, 43, 56, 58, 61, 109
sisu, 127
Slim, William (Field Marshal Lord), quoted, on originality, 127
small groups, and battle performance, 156
"small wars," 167; as fractalization, 7, 22; in New World Order, 168
Smith, Dale O., (Brig.Gen.), quoted, on doctrine, 37
Smith, Walter Bedell (Lt.-Gen.), quoted, on fear, 101
social classes tensions, and military organization, 159; and secrecy, 172
soldiers' battles, 36, 124
soldiers, images of, 82–83
Solferino, Battle of (1859), 9
sole authorship, in writing of academic history, 145
Somme campaign on (1916), 6; command problems in, 89; German withdrawal along, 1917, 140
Soncrant, Roy, on historical models, 110 n. 5
SOPs (standard operating procedures), as form of doctrine, 28, 35, 104
Sorpe dam, RAF attack on, 1943, 140
South Vietnam, forces of, relations with U.S. forces in Vietnam War, 137
Soviet Union 8, 15; armed forces doctrines, 91. *See also* Red Army; Stalin, Joseph; World War II
Spaatz, Carl (Gen. of the Army), 61, 67
Spain, guerrilla war in (1807–14), 7
Spanish Civil War (1936–39), unescorted bomber losses in, 43, 56, 58, 109
Spanish-American War (1898), 47
Spanish Foreign Legion, 53
Special Air Service, in North Africa, 1941–42, 109
special military units, 93, 101
Special Operations, 151
Speer, Albert (Min. of Armaments), reorganizes German war production, 67
Spruance, Raymond F. (Adm.), command style of, 83
"squad-leader-in-the-sky" syndrome, in Vietnam war, 81, 178
"square law," of fire exchange, 26
staff officers, historical obscurity of, 120
Stalin, Josef [Djugashvili] (Premier), 41, 43; copes with chaos, 1941–42, 166
Stalingrad, Battle of (1942–43), 20, 39; chaos described in, 91; communications in, 81
state changes, in crisis and war, 122–124
statistical mechanics, 26, 74
statistics, of casualties, in Gulf War, 173
Steiner, Felix (General), theories of, 90
stochastic patterns, as elements of war, 145
stormtroops, World War I, 53, 79, 101
strategic bombing, in World War II, 47, 58, 67, 68, 90, 99, 135, 159; in Korea, 109; in Vietnam War, 135
Strategic Bombing Survey. See United States, Strategic Bombing Survey
strategic intelligence, varying definitions of, 147
Strauch, Ralph, on perception-action link, 70
Streicher, Julius, mental state of, 161
The Structure of Scientific Revolutions, 11, 106
style, of military leaders: and creativity, 115–116; as source of chaos-complexity, 69 passim; of officer corps, 98, 160, 173; various commanders', 80, 82, 83, 98, 108
submarine warfare: and international law, 161; as fractalization of tactics, 6–7, 100, 101; in World War I, 108
subtle forces, disproportionate influence of: and cumulative environmental damage, 135–143; in history, 172; in intelligence analysis, 149; in military history, 21, 92; in military operations, 11, 24, 25, 79, 81, 82, 98, 107, 159, 174
Suez affair (1956), leaders' health in, 161

Sumner, William Graham, quoted, 3
Sun Pin (Gen.), quoted on coping with confusion, 90
superpower rivalry, aftermath of, 166–167, 174–175; constraints on use of power, in Cold War, 167 polarizing effects of, 178
surprise attacks, 10, 44; and creativity, 126
symmetry, and casualties, 7; in battle dynamics, 153; in Gulf War, 160; in science and philosophy, 169; in tactical formats, 5–6, 10
synergies, in war, due to collision of forces, 126
systems: analysis of, 97; analysts, 117; theories, 11, 170

tactical control, in Gulf War, 173
tactical formations, symmetry of, 5, 6, 10, 109, 153: fractalization and, 54–57, 99; in Boer War, 80, in Gulf War, 160. *See also* battle dynamics; command and control; combat; leadership; tactics; technology; war; warfare
tactics, abstracted in theory and history, 125; chaos-complexity and, 178; graphic portrayal of, 105; linearity and symmetry in, 153
Tai'ping Rebellion, late 19th Century, 41
Tamerlane, 116
Tank, Kurt, German aircraft designer, 60
Tank destroyer Corps, U.S. Army, World War II, 54–56
tank advocates, adherence to mass, 54–56
tank destroyers, in World War II, 53–56; lack of "friend at court," 69
tanks: as tactical surprise, World War I, 68, 132 n. 5; initial surprise potential of lost, 68; concepts derived from Mongols, 47
Tarawa, assault landing on (1943), 125
Taylor, Frederick Winslow, industrial efficiency concept of, 169
Taylor, Zachary (Maj.-Gen./ Pres.), command style of, 102
TECHINT (technical means of intelligence gathering), 146; infatuation with, 1970s, 151
technology: and chaos in war, 69, 88, 93; and collateral damage, 99; and rise of dictators, 84; effect of, on discipline, 78; documentation of war and, 173; on environment, 141; on history, 16, 21, 23; on naval tactics, 50; impact on war, 8, 23, 97, 100, 156, 177
telegraphy, effect of, on warfare, 8, 177
telephone, effect on documentary evidence of history, 23
television, battle images on, 21; in Gulf War War, 173
Teplov, Boris (Maj.-Gen.), on operational thinking, 1307
terrorism, chaos as goal of, 93
theoretical physics, history of, 131
thermonuclearetics, 117
"thinking bayonets," 80
Third Reich, armored forces of, 55
Thirty Years' War (1618–48), atrocities in, 99
"timeless verities of combat," 104
"toggleiers," 65
Tojo, Hideki (Gen.), strategic gamble of, 1941, 40
Tolstoi, Leo, 22, 30, 166; quoted, on view of war, 75
Totalenkrieg, 21
total war, 99
trade policy, U.S., lack of, in Cold War, 153
trajectories, of events, in war, 176
"trench cleaner," 101
trench mortars, 101
Trenchard, Hugh (Air Chief Marshal), 28
tribal warfare, 109, 129
"Troubles," Irish (1916–22), 7
Truppenfuhrung, German Field Service Regulations (1933), recognizes chaos of battle, 5
Tunisian campaign (1942–43), 20, 55
Tupolev, Andrei, Soviet aircraft designer, 61
turbulence, 1; in military theory, 166; of combat, 48–49; in warfare, 48–49; psychological, 81
Turenne, Henri de la Tour d'Auvergne (Marshal/ Viscount), on relative error in battle, 40

29th Infantry Brigade, U.S. Army, 53
twin attractors, in heavy-light forces debates, 49
"two up and one back," tactical maxim, 104

U-boats, surprise potential of lost, in World War I, 68; losses of, World War II, 23, 65; successes, with German Air Force, in World War II, 100
Ukraine, flooding of, 1941, 141
uncertainty, and defense planning, 174; in physics, 177; of historic evidence of warfare, 1, 34; principle, Heisenberg's, 130
uniforms, military, pattern since French revolution, 152 n. 6
Union Army, American Civil War, casualties of, 9
United States:
Air Force, 62
Air Service, 43, 56
Army, Command and General Staff College, 58; General Staff, War Plans Division, 58; growth, World War II, 103; in Gulf War, 173; in Korean War, 106; in Vietnam War, 134–138; morale concerns in, 125, 155; "fraggings" in Vietnam War, 160; Rangers of, 53; Tank Destroyer Corps of, 54–56; 29th Infantry Brigade of, 53; weapons systems growth in, 167
Army Air Corps, G.H.Q. Air Force, 77; escort fighter problem in, 56 passim
Army Air Corps Tactical School, 57, 59
Army Air Forces, 83, 84, 87, 95, 96; escort fighter problem in, 61–62, 63, 64–67
Congress, intelligence hearings, 1975, 151
Constitution of, 161, 171
Marine Corps, on Tarawa, 123; "Old Breed" of decimated, 13; demeanor of officers and NCOs, leaders' demeanor, 160
Military Academy, engineer training at, 138
Navy, aircraft carriers in, 108; after Pearl Harbor, 35; at Pearl Harbor, 7; Maritime Strategy of, 70; P.T. boats of, in World War II, 51–52
Strategic Air Forces, Europe, 67
Strategic Bombing Survey, 89

Van Creveld, Martin, on low-intensity conflict, 168
Vagts, Alfred, 38
vectorial laws, 18
Vegesack, 8th Air Force raids on, 64
Verdun, Battle of (1916), 47, 93; soldiers' battles in, 124
Versailles Treaty (1919), 108
veterans, perspectives of war, 87; of Gulf War, mysterious ailments of, 143; views of military history, 2, 21, 30, 77, 157
Vichy France, 60
victory-defeat dynamic, 20, 78, 91, 103, 123, 124, 163–164; "what-iffing," and, 171
Vietnam War (1961–75), 104, 156, 172; and Munich precedent, 47; bombing in, 135, 137; complexity of, 137–138; dearth of baseline data in, 117; environmental warfare in, 135–139; effect on U.S. defense structure, 98; cohesion and bonding concern, following, 125; "fraggings" in, 160; numerical analysis in, 138; opposing leaders' health in, 161; constraints on U.S. use of force in, 167; systems analysis confounded, 98; U.S. junior officers' resentment of command practices, 178
Viking raids, 116, 129
violence, human, impulse to, 13; roots of, 157; sexual constraints and, 158
virtual reality, 178
Volga river, 157
von Clausewitz, Karl (Maj.-Gen.), on simplicity in war, 107; on war as "province of chance," 32, 106; on war as instrument of policy, 162; Soviet view of, 166
von Lettow-Vorbeck, Paul Emil (Gen.), East African campaigns of, 1914–18, 7
von Karman, Theodore, quoted, on theoretical testing limits, 178
von Ludendorff, Erich (Gen.), out of phase with reality in 1918 campaigns, 106
von Manstein, Erich (Field Marshal), craftiness of, 120

von Moltke, Helmuth Karl Bernhard (the Elder), defines strategy, 174; on fragility of plans, 37
von Rennenkampf, Pavel Karlevich (Gen.), rivalry with Samsonov, 1914, 70
von Richtofen, Wolfram (Gen.), on end of *blitzkrieg* phase in World War II, 90
von Schlieffen, Alfred (Field Marshal Count), theories of, 9, 28, 31, 46, 104
von Seeckt, Hans (Gen.), on impermanent nature of armies, 103
V-1 (*Vergelgtungswaffe* [vengeance weapon] 1), Nazi cruise missile, 61
Voyennaya Mysl', 93
vyostny istrebitel', 61
vyostny perekvatchik, 61

Wagner, Richard, 140
Walsingham, Francis (Sir), 152
war, war dynamics, warfare: amateurs' success in, 162; as art-form, 115–118, 123; as collision of forces, 166; as pathology, 97–98; as "province of chance," 32, 106; as struggle to shape future, 41; causation and, 19, 36; chaotic form of, 5, 12, 32, 81, 93, 99, 176; creativity and, 114–131; damage of, 135–143; disorder of, 24; economic paradoxes of, 153; emotional aspects of, 131–133, 161–163; environmental effects of, 97, 135–143; evidence of, 1, 21; experiences of, 157; fuzziness of, 168; images of, filtered, 78, 97; metaphors of, 124–127; non-linear perspectives on, 106, 145–148; momentum of, 110 n. 5; reveals elites' inadequacy, 170; ritualization of, 128–129; Romans' view of, as chaos, 94; sexuality and, 155; statistics of, 23; terminal chaos in 178; technology, effect on, 8; turbulence in, 48, 50, 99; types of, in cold war, 167; unpredictability of, 5
war plans, and creativity, 124–130; World War I, 46–47
War Plans Division, U.S. Army General Staff, 58
"war within the war," German Army, World War I, 69
"warrior spirit," 80
War of National Liberation, 167
War Plans Division, U.S. Army General Staff, 50
Washington, D.C., 138, 139
Washington, George (Lt.-Gen.), at Germantown, 16, 128
Waterloo, Battle of (1815), 6, 20, 25
weather, effects of, on war, 162
Webster's New International Dictionary, 2nd ed., doctrine definition, 35
Wehrmacht, assault guns of, 55; doctrine of, and chaos, 94; in Russia, World War II, 47, 90, 91, 139, 165
Wellington, [Wellesley, Arthur] Duke of, 2; quoted, on battle dynamics, 29; on military history, 91–92; on Waterloo, 20, 25; victories of, 164
Westover, Oscar (Maj.-Gen.), 57
West Point. See United States Military Academy
"what-iffing", 23, 171–172
Whaley, Barton, on surprise attacks, 126
white phosphorous shells, use of, in World War II, 38, 161
"whiz kids," Pentagon systems analysts, 1960s, 31, 117, 137
Wilderness campaign (1864), chaos in, 124
Wilhelmshaven, 8th Air Force raids on, 63, 64
Willoughby, Charles (Maj.-Gen.), misappreciations of, in Korean War, 44; in World War I, 164
Wilson, Woodrow (Pres.), as failed visionary, 43; as peacemaker, 162
Winchester, Battle of (1864), 20
Winter War, Finland (1939–40), 39
Wohlstetter, Roberta, 148
Wood, John (Maj. Gen.), armored warfare precepts of, 104; on need for range of styles, 115
World War I, 6, 7, 10, 41–47, 89, 93, 99, 100, 107, 124, 137, 154; air tactics in, 109; command process in, 87; environmental damage in, 139–140; fractalization of tactics in, 53; French Army; leaders' styles in, 83; mutinies in, 78;

Napoleonic models in, 121; naval warfare in, 50–52; tactics in tactical surprises in, 93; trench warfare tactics in, 104, 108
World War II, 10, 17, 23, 39, 41, 77, 88, 89, 98, 99, 107, 125; airborne forces, 101–102; amphibious operations in, 90, 102; applied science in, 117; leaders' health in, 161; light forces in, 53; massing and dispersion in, 90; origins of, 178; personnel pipeline problem in, 103; U.S. force types in, 167; venereal disease in, 150
Wright brothers, pacific hopes of, 131
Wright Field, 62
Wylie, J. C. (Capt.), quoted, on limits of predictability, 117

Yamamoto, Isoroku (Adm.), "assassination" of 1943, 22, 80
Yamashita, Tomojuki (Gen.), 106
Yamashita, Yosimichi, quoted, on managing creativity, 119
York, Alvin (Staff Sgt.), 157
Young, Rodger (Pvt.), 157
Ypres, Third Battle of, flooding in, 1917, 213. *See also* Passchendaele
Yugoslavia, devolution of, 158

Zeitzler, Kurt (Col. Gen.), quote, on Stalingrad chaos, 91

About the Author

ROGER BEAUMONT has taught history at Texas A&M since 1974. He served two tours of active duty with the army as a military police officer. A cofounder and former North American editor of *Defense Analysis,* Beaumont was the first historian named as a Secretary of the Navy Fellow at the U.S. Naval Academy. The most recent of his ten books and monographs is *Joint Military Operations: A Short History* (Greenwood, 1993).

ISBN 0-275-94949-4
90000>
EAN
9 780275 949495
HARDCOVER BAR CODE